The Street Names of England

Other books on names by Adrian Room

Place-Names of the World (David & Charles, 1974)
Place-Name Changes Since 1900 (Scarecrow Press, 1979)
Naming Names (McFarland, 1981)
Dictionary of Trade Name Origins (RKP, 1982)
A Concise Dictionary of Modern Place-Names in Great Britain and Ireland (OUP, 1983)
Guide to British Place-Names (Longman, 1983)
Dictionary of Translated Names and Titles (RKP, 1985)
Dictionary of Irish Place-Names (Appletree Press, 1986)
Dictionary of Coin Names (RKP, 1987)
Place-Names of the World (Angus & Robertson, 1987)
Dictionary of Astronomical Names (Routledge, 1988)
Dictionary of Place-Names in the British Isles (Bloomsbury, 1988)
Dictionary of World Place-Names Derived from British Names (Routledge, 1989)
A Dictionary of Pseudonyms and Their Origins (McFarland, 1989/St James, 1990)
NTC's Dictionary of Trade Name Origins (National Textbook Company, 1991)
Corporate Eponymy (McFarland, 1992)
A Name for Your Baby (Marboro Books, 1992)
Brewer's Dictionary of Names (Cassell, 1992)
Place-Name Changes 1900-1991 (Scarecrow Press, 1992)
A Dictionary of First Names (Marboro Books, 1993)
The Naming of Animals (McFarland. 1993).

THE STREET NAMES OF ENGLAND

by

ADRIAN ROOM

PAUL WATKINS
STAMFORD

1992

Photoset from the discs of the author
and published by
Paul Watkins
18, Adelaide Street,
Stamford,
Lincolnshire,
PE9 2EN.

ISBN
1 871615 20 8

Printed on long-life paper

Printed by Woolnoughs of Irthlingborough

They explored every corner of the town. It was a place of pleasant sounding old names, richly English and romantic. [...] The names of the streets fascinated Christopher: Green End, Lombard Street, Baileygate, Golden Hill, the Tything, Market Row, Vine Court, Barbican, Angel Alley.

Warwick Deeping, *Sorrell and Son*,
Chapter VII, Part 3 (1925)

ACKNOWLEDGMENTS

I owe thanks to all those who have approved of and encouraged this venture, and especially Shaun Tyas, of Paul Watkins Publishers, who commissioned the book in the first place and who was more than patient when involvement with other work obliged me to delay its delivery to him for far too long. As both publisher and editor, he has shown unfailing enthusiasm and interest in the project, and I hope that he, and readers of the book, will find something of the reward and enjoyment in the final product that I myself experienced in writing it.

My thanks are also due to John Field, who read the book at proof stage and from his expert knowledge of the subject made a number of helpful comments.

Since a book is also sometimes judged by its cover, I would in addition like to thank Celia Tyler for the latter's elegant and eloquent pictorial design.

Several of the individual names in the book, especially the historic ones, were drawn from the county volumes of the English Place-Name Society, and I am grateful to Professor Kenneth Cameron, Hon. Director of the Society, for kindly permitting me to quote from these works as appropriate.

Adrian Room
High Street, St Martin's, Stamford
August 1992

CONTENTS

ABBREVIATIONS

The following are the abbreviations used for English counties in this book:

Avn	Avon	Ken	Kent
Bed	Bedfordshire	Lan	Lancashire
Ber	Berkshire	Lei	Leicestershire
Buc	Buckinghamshire	Lin	Lincolnshire
Cam	Cambridgeshire	Mer	Merseyside
Che	Cheshire	Nor	Norfolk
Cle	Cleveland	Not	Nottinghamshire
Cor	Cornwall	Ntb	Northumberland
Cum	Cumbria	Ntp	Northamptonshire
Der	Derbyshire	NYo	North Yorkshire
Dev	Devon	Oxf	Oxfordshire
Dor	Dorset	Shr	Shropshire
Dur	Durham	Som	Somerset
Ess	Essex	Sta	Staffordshire
ESx	East Sussex	Suf	Suffolk
Glo	Gloucestershire	Sur	Surrey
GtL	Greater London	SYo	South Yorkshire
GtM	Greater Manchester	T&W	Tyne and Wear
Ham	Hampshire	War	Warwickshire
H&W	Hereford and Worcester	Wil	Wiltshire
Her	Hertfordshire	WMd	West Midlands
Hum	Humberside	WSx	West Sussex
IoW	Isle of Wight	WYo	West Yorkshire

All places are followed by the abbreviated name of their county with the exception (for brevity's sake) of the following towns and cities, all of which but for three are in counties named for them: Bedford, Birmingham, Bristol, Buckingham, Cambridge, Chester, Derby, Durham, Gloucester, Hereford, Hertford, Lancaster, Leicester, Lincoln, Manchester, Northampton, Norwich, Nottingham, Oxford, Stafford, Warwick, Worcester, York. (Strictly speaking, Shrewsbury also belongs here, but the etymological link between the town's name and that of its county, Shropshire, is not always readily made. The same certainly applies to Southampton and Hampshire.) The three 'odd men out' are Birmingham (WMd), Bristol (Avn) and Norwich (Nor). The first of these is the largest city in England after London and hardly needs its county name. The second was long a county in its own right. The third, perhaps through the (false) association between its name and that of its county, is equally familiar. The general location of each (West Midlands, West Country, East Anglia) is common knowledge.

'London', in the main body of the book, means Greater London, an administrative region often larger than many suppose, and taking in such well-known towns (some even now associated with their historic counties) as Barking, Bromley, Croydon, Enfield, Harrow, Hornchurch, Kingston upon Thames, Orpington, Richmond, Romford and Uxbridge.

INTRODUCTION

Everyone notices street names, if only because they figure so largely in our daily lives, whether in addresses or in destinations. But where do the streets actually get their names?

Take a town plan at random, say for Nuneaton, in Warwickshire. We see the town has *Abbey Green*, *Alexandra Street*, *Back Street*, *Bond Street*, *Bottril Street*, *Cheverel Street*, *Church Street*, *Clinic Road*, *Dugdale Street*, *Fife Street*, *Glebe Road*, *Henry Street*, *High Street*, *John Street*, *Kingsway*, *Leicester Road*, *Market Place*, *Meadow Street*, *New Town Road*, *Oaston Road*, *Park Street*, *Queen's Road*, *Riversley Road*, *Roanne Ringway*, *Rose Lane*, *Simon Close*, *Victoria Street*, *Weddington Road* and *Wheat Street*, among many others. Where do they all come from?

We may be able to make an informed guess about some. *Alexandra Street* looks like a royal name, as do *Kingsway*, *Queen's Road* and *Victoria Street*. *High Street* and *Market Place* are familiar enough, and doubtless *Leicester Road* leads to Leicester. But does *Oaston Road* lead to Oaston, or is that a personal name? Is *Riversley* a place or a person? If a place, what sort of a place? If a person, who was he or she? Who was the Simon who gave the name of *Simon Close*, and more unusually the Roanne who had a ring road named for her? Was wheat sold in *Wheat Street*, or was there a pub here, called perhaps the *Wheatsheaf*? How does *Fife* come into it, all these miles from Scotland?

Alas, only local knowledge will reveal the truth behind the vast majority of these names. And the same applies in other towns. True, there are the regularly occurring *High Street*, *Church Street*, *Market Place* and so on. But for the rest it is no use guessing. We cannot sometimes even be sure that we have written or remembered an unfamiliar street name correctly. One plan of Nuneaton (let us spare its publisher's blushes) prints the strange-looking *Roanne Ringway* as the two separate streets of *Roxanne* (*sic*) and *Ringway*, and a hotel in another town in a street named *Weston Road* regularly receives mail addressed to *Western Road*.

INTRODUCTION

Clearly streets and roads actually need to be named, just as towns and villages do. But what is their name to be?

Streets have been named ever since towns have been built, and it follows that the oldest towns will therefore have the oldest names. Some such names are as old as the English language, and have come down to us in corrupt form. Many medieval names are now distortions of their original form. York's *Coney Street* is nothing to do with rabbits but means 'king's street'. (For the sense and source of the name, see Chapter 3, p. 36.) Norwich's *Charing Cross* has deviously evolved from a name that in medieval times meant 'shearer's hill' (*Scherershil* in a document of 1309) and that was originally *Tonsoria*, from the Medieval Latin word for a workshop where woollen cloth was shorn.

Such names represent a way of life quite different from our own, and are all the more valuable and interesting for that.

The nature or even history of a town is to a greater or lesser degree encapsulated in its street names. Old cities like York and Norwich contain a fair number of medieval names. But what sort of a town is it that has names as follows: *Appleyard Place*, *Booker Avenue*, *Cairngorm Gate*, *Conniburrow Boulevard*, *Fennel Drive*, *Hampstead Gate*, *Lower Ninth Street*, *North Eleventh Street*, *Portway*, *Springfield Boulevard*, *Wandsworth Place* and *Witan Gate*? This seems to be an eccentric, even eclectic combination of old rural names and modern urban ones. In a sense this assumption is correct, for the town that accommodates these names is none other than Milton Keynes, laid out as a New Town in the 1970s on a virgin site by a small village of the name in Buckinghamshire. The town was planned carefully, and the names of its streets were designed as carefully as its buildings. The city is laid out on a rectangular plan, with tree-lined streets named *Boulevard* running from east to west, and shorter streets named *Gate* running from north to south and intersecting them. Even shorter streets running from one boulevard to another are given ordinal numeral names and called *Street* and *North* or *South* appropriately. The rural names are those in the still verdant residential area outside Central Milton Keynes. *Conniburrow* is a genuinely old field name, meaning what it (for once) seems to mean: 'land where rabbits live'.

Harlow New Town, in Essex, was laid out in the 1950s and 60s and was given almost exclusively rural names, although there is the odd *Hammarskjold Way* and *Mandela Avenue* in among the *Greenhills*, *Parsonage Leys*, *Ryecroft Wrens* and *Westfield*. But the presence of the former gives the lie to the latter. The town may have a rural aspect and location, but it is newly planned and built. One can even date it: Dag Hammarskjold was general secretary of the United Nations from 1953 to 1961, and his premature death in an air crash made it all the more desirable that his name should be commemorated in some way. Nelson Mandela, the famous black South African political leader, was jailed in 1962 for five years and in 1964 for life. He, too, deserved his tribute. The lives and deeds of both men are commemorated in the names of the streets that were laid out at the time when they were in the news.

One more 'portrait'. Here are some streets of Nelson in Lancashire, a textile-manufacturing town that arose in the late 18th century and that developed in the 19th: *Albert Street*, *Arthur Street*, *Baker Street*, *Belgrave Street*, *Chapel Street*, *Commercial Road*, *Cuba Street*, *Derby Street*, *Edith Street*, *Elizabeth Street*, *Every Street*, *Fountain Street*... One can see the pattern and, frankly, the unoriginality or at any rate uniformity of the names, the *Streets* easily outnumbering the *Roads*.

There is or was an important difference between *street* and *road*, and the first chapter of this new study of street names is devoted to it. It is thus also a study of road names.

Street names can be approached from a number of angles. One can see them as derivative or functional. They can be categorised by origin or by type. Inevitably, some of the categories overlap and blend. *St John's Road* is both a religious name and the name of a street that leads to St John's church. Streets can have names of local or distant reference. *London Road* is usually the name of a road that leads out of a town to London.

Northampton Square (say) in a town nowhere near Northampton is likely to be named for a person, or more precisely for an aristocratic title. The connection between the named person and the square is one that only local archives will reveal.

In the very nature of things, streets are extended, rebuilt, parcelled out, even razed from the face of the earth in some planning scheme, perhaps one of slum clearance or commercial development. This means that street names, more than any other type of place-name, are themselves subject to disturbance and disappearance. Many streets in many towns have been renamed. In Norwich, for example, just mentioned, *Bank Street* was *Blue Boar Lane*, *Brigg Street* was first *Over Wastelgate* then *Market Lane*, *Horns Lane* was *Skeythegate*, then *St Bartholomews Lane*, then *Adcock Lane*, and *Rigby's Court*, named for an 18th-century doctor, was originally *Pit Lane*, for a local refuse pit. 'Improvements' like this are not uncommon in the process of street-naming. Meanwhile, *Elmeswell Lane*, *Griffin Lane*, *Long Lane*, *Ratton Row* and *Synagogue Street* are just some of the Norwich names that have disappeared altogether from the map. The last-named here was destroyed at a stroke during a German air-raid in 1942.

This book is almost entirely concerned with the names that exist today. Inevitably, some former and historic names will be mentioned, but it is the current names that stimulate the greatest interest, simply because they are those of here and now. Many of them, such as London's *Oxford Street* and *Trafalgar Square*, are world-famous. Others are familiar only locally. But they all have a tale to tell, and this book tells it, as far as it is known.

It will be appreciated that, because the origins of most street names are purely local, there are hundreds of streets whose names will not feature here. They were not included in the survey. On the other hand, hundreds of streets *are* represented here, and their representation is sufficiently large to enable the history of street names and the whole process of street-naming to be examined in some depth.

A word about geographical location. Streets are located by town and towns by their modern county (see the Abbreviations for these, p. viii). In Greater London, streets may be identified by district or region as well as by 'compass point', that is, in north London, southwest London, and so on. For the purposes of the latter specification, 'north', 'south' and the like relate to the City of London. This means that 'south' will be across the Thames, even if only *just* across, in Southwark. A compass description does not relate to distance, therefore, so that Westminster is 'west London', as it is west of the City (hence its name, in fact), as also is Uxbridge, formerly (or even now, depending on one's view of these things) in Middlesex, but officially in Greater London and a good 16 miles west of the City. But enough geographical information is given to make the location of the particular street reasonably easy to identify, even if a map of London is needed. In the final analysis, too, it is the name itself that is important, and the relevance of its particular location, where it exists, will be included in the account of its origin.

1. STREET OR ROAD?

The two most common generic terms for a roadway are street and road. What is the difference?

The first and obvious distinction is that, in general, a *street* is usually found in a town, while a *road* is in the country. The second difference is that a *street* is basically a place of location, where people live or work, whereas a *road* normally leads to a specific place. For this reason, a road is often named for its destination, such as London's *Edgware Road*, *Old Kent Road* and *Harrow Road*. For the same reason, a road is normally much longer than a street.

A third and more subtle distinction is that, in spoken English, the name itself is stressed in a street name, but in a road name both parts of the name are equally stressed. Two obvious examples are London's *Oxford Street* (where 'Oxford' is stressed) and *Edgware Road* (where 'Edgware' and 'Road' are equally stressed). Moreover, in colloquial use, a road name is frequently prefixed by *the*, as in *the Edgware Road* and *the Old Kent Road*.

This third distinction springs from the historic origin of the respective words *street* and *road*. *Street* is one of the few loan words adopted early into English from Latin. (Others are *mint*, from *moneta*, *pound*, from *pondus*, and *sack*, from *saccus*.) During their occupation of Britain, from the 1st century BC to the 5th century AD, the Romans referred to a military highway, one they had themselves built, as a *strata*, a word that itself evolved from Classical Latin *via strata*, literally 'paved way'. When the Anglo-Saxons came to Britain in the 5th century, they adopted this word for the existing Roman roads, which they had not seen before, and for which they had no native name. Their word *strǣt*, which eventually gave modern *street*, was subsequently applied to any road or route. When following a specific word or describer, as in *High Street* or *Devonshire Street*, the first word is stressed for purposes of differentiation. A similar stress pattern is found for the names of rivers, such as the

River *Thames*, although here (at least in British usage) the name follows the generic term.

Road, on the other hand, although a native Anglo-Saxon word (Old English *rād*), did not gain its present meaning until the 16th century, when it was promoted by Shakespeare and his contemporaries. Its original sense, recorded from the 9th century, was *riding* (the two words are directly related). It thus came to be associated with a specific journeying on horseback between two places. Looked at another way, a *road* at first denoted an impermanent action, a single journey, whereas a *street* was a permanent paved route. The word *road* evolved its present sense only when a regular route between two places had become as permanently established as a Roman road was, even if its surface was still largely 'unmade'.

Significantly, we still speak generally of Roman *roads* (the Anglo-Saxons *rode* along them), but name them as *streets*, preserving the modern equivalent of the Latin word, as for *Ermine Street* or *Watling Street*. It should be observed that in general the actual names of the Roman roads themselves are Anglo-Saxon. (A noted exception is the name of the *Fosse Way*, where the first word represents Latin *fossa*, 'ditch', with reference to the ditch that runs on either side of it.) If the Romans themselves named their roads, the names have not come down to us.

The implicit 'action' sense of *road* resulted in the word receiving equal stress with any accompanying name or adjective. A similar level stress is found in phrases containing its modern equivalent, *ride* or *journey*, so that we talk of a '*long ride*' or a '*tiring journey*'.

Today, most towns and cities boast a number of generic terms for different types of thoroughfare, with *alley*, *avenue*, *close*, *grove*, *hill*, *lane*, *row*, *walk* and *way* just some of those encountered in addition to *street* and *road*. Few towns illustrate the range better than Kendal (Cum), which has capitalised on its ruined castle to create the modern street names *Castle Circle*, *Castle Close*, *Castle Crescent*, *Castle Dale*, *Castle Drive*, *Castle*

Garth, *Castle Grove*, *Castle Oval*, *Castle Park*, *Castle Riggs*, *Castle Rise*, *Castle Road*, *Castle Street* and *Castle Walk*, among others, with *Castle Green Lane* a stretch of the A684 that runs south from the town to Sedbergh (Cum). (For further names of this type, see Chapter 13, p. 179.)

It would therefore be useful to say something here about the generic terms that are most commonly found in street and road names, giving their definitions and quoting examples. It should be pointed out, however, that in some cases there appears to be little significant difference in modern usage between one term and another. In certain instances, too, the selected term is virtually arbitrary or even meaningless. It does not therefore follow that *Manor Gardens* is noted for its horticultural display, that *Acacia Avenue* is lined with such trees (it may have no trees at all), or that *Pinewood Close* is an enclosure, let alone one in or near pine woods. Even so, some such designations have retained a little of their original sense, even where this is historic or archaic.

alley An *alley* was originally a walk bordered with trees or shrubs, typically in a park, garden or wood. In a modern urban setting, however, the term is frequently used for a narrow lane between buildings or walls, usually for pedestrians only. *Pepper Alley*, Epping Forest (Ess), is an example of the rural application, while *Great Swan Alley*, in the City of London, represents the urban usage. The word itself is related to French *aller*, 'to go', and implies a route for walking or riding.

approach A road or street that gives access to a place, now especially a railway station. The term became popular from the 1870s, and is typically found in *Station Approach* in several towns. The name may still exist in cases where the station has long been closed.

arcade The term usually designates an enclosed or semi-enclosed row of shops. London's *Burlington Arcade* is fashionably familiar. Other examples are *The Arcade*, Barnsley (SYo),

Homewell Arcade, Havant (Ham), and *Queens Arcade*, Nuneaton (War).

avenue The term was adopted in the 17th century from the French, in which *avenue* (from a word related to modern French *avenir*, 'to come to') was a tree-lined route. In English use an *avenue* was thus originally a similar approach to a country house. The word later came to apply to any broad way lined with trees or other objects. The street name *The Avenue*, without further addition, is widely found (see Appendix, p. 206), while London's *Shaftesbury Avenue* is one of the best known individual examples. The French origin of the word can make it appropriate for a modern street name based on a English town's French 'twin', such as *Avenue de Chartres*, Chichester (WSx), the name having a French form overall.

bank As the term implies, the application is often to a road or street that rises up (or descends down) a slope, or that is on relatively high ground by a river or stream. Examples are *Anchorwood Bank*, Barnstaple (Dev), *Church Bank*, Bradford (WYo), *Kilner Bank*, Huddersfield (WYo) and *Roman Bank*, Stamford (Lin). A street of the name may have been built along the bank of a river, as London's *Millbank*.

boulevard The term denotes a wide, usually tree-lined road, and was adopted from the *boulevards* of Paris and other French cities in the 19th century. The French term itself derives from the Dutch word for *bulwark*, as such roads were originally built along ramparts. In modern English use, the term is favoured for through routes and relief roads, especially in new towns or suburban residential developments. One example is *Bourges Boulevard*, Peterborough (Cam), a section of the A15 appropriately named for the city's French twin town (see Chapter 12, p. 167).

brow The term usually applies to a hill or rise, with the word itself related to Scottish *brae*. Examples are *Everton Brow*, Liverpool (Mer), and *Pinners Brow* and *School Brow*, both in Warrington (Che).

buildings The word is used for a short street leading past a group of buildings, or a tenement block, usually in a dense city area, with the name usually that of the buildings themselves. Examples are *Copthall Buildings*, *Dorset Buildings* and *Royal Exchange Buildings* in the City of London.

causeway The word, deriving ultimately from Medieval Latin *via calciata*, 'way paved with limestone', is now chiefly used for a raised road across marshy or low-lying ground. The fenland city of Cambridge thus has its *Maids Causeway* while a well-known London example is *Newington Causeway.*

circus The name is usually found for a circular range of buildings. London's *Oxford Circus* and *Piccadilly Circus* are well-known examples, and *The Circus*, Bath (Avn), is architecturally famous. Instances elsewhere include *Dartmouth Circus*, Birmingham, *Drakes Circus*, Plymouth (Dev) and *Pitville Circus*, Cheltenham (Glo).

close As it implies, the word originally designated an *enclosed* area, such as a medieval courtyard. By extension, the term came to be used for the entrance to such a courtyard. In modern usage, the word is often found for a quiet residential road, frequently one with no outlet (ie a cul-de-sac). *Church Close* is common, and London has its *Neville Close*, *Park Close* and *Sydney Close*, among many others.

cottages The name denotes a short row of cottages, and so a street or alley that runs by such a row. Examples are *Albert Cottages*, Falmouth (Cor), *Prince Consort Cottages*, Windsor (Ber), *Shaftesbury Cottages*, Plymouth (Dev), and *Zebra Cottages*, Stamford (Lin).

court The basic sense is that of a courtyard, as typically in a castle. The term later came to apply to a yard that opened off a street, and by extension to the short passage that led to such a yard. Examples of the latter can be found off London's Fleet Street: *Red Lion Court*, *Bolt Court*, *Poppins Court*. In modern times the term frequently applies to a block of flats, often with a well-known place- or personal name, such as *Hastings Court*,

Selwyn Court. This later usage developed from the adoption of the word for the names of country houses, such as *Hampton Court*, *Squerrys Court*. For the purposes of this book, *court* meaning 'block of flats' counts as a street name.

crescent A semi-circular or curving street, as the descriptive term implies. The name was originally that of a line of houses built in such a curve. Bath's famous *Royal Crescent* dates from 1784, and was the first street in the country to be named with this word.

croft The word has not altered since Anglo-Saxon times, when it was used of a small enclosed field. In the north of England it specifically applied to such a field by a house. In street names it now has something of the latter sense, as if referring to a row of houses and their individual gardens. Harlow New Town (Ess) has *Oldhouse Croft*, *Orchard Croft* and *Sharpe Croft*, among others, but the name is also found in older towns, even if on modern estates, as for *Goose Croft*, *Toms Croft* and *Well Croft*, Hemel Hempstead (Her).

drive The original sense of the term was for the private road that led to a large house, especially when the route was tree-lined or 'scenic'. In modern times the designation is frequently given to one of the longer streets in a suburban built-up area, especially one that leads to a main road. Examples of the latter include *Denbigh Drive*, Fareham (Ham), *Buchanan Drive*, Luton (Bed), *Pemberton Drive*, Bradford (WYo), and *Tiverton Drive*, Nuneaton (War). Ipswich (Suf) has *Civic Drive* as one of its modern central streets. Compare *drift* and *drove*, p. 15, and see also the Appendix, p. 193.

embankment The name specifically applies to a road built along a bank that has itself been constructed to contain a river. Two of the most familiar examples are London's *Albert Embankment* and *Victoria Embankment*.

esplanade The word is French in origin, ultimately deriving from Italian *spianare*, 'to make level'. In English use it traditionally applies to a type of seaside promenade, a road along the sea front suitable for walking, strolling or sitting. Its application is mostly restricted to seaside resorts or waterside sites, and this in general distinguishes the term from *promenade*, which can be found in various types of location. The name usually occurs simply as *The Esplanade* and is found in such seaside resorts as Bognor Regis (WSx), Bridlington (Hum), Cowes (IoW), Dover (Ken), Penzance (Cor), Poole (Dor), Ramsgate (Ken), Ryde (IoW), Southend-on-Sea (Ess) (as *Eastern Esplanade* and *Western Esplanade*), Southport (Mer), Weymouth (Dor), and Worthing (WSx). Some resorts, such as Bognor Regis, have both an esplanade and a promenade. Inland, York has *The Esplanade* as a riverside road, while Rochdale (GtM) also has *The Esplanade* as one of its most important central streets.

gardens The name is now loosely found for any modern residential street, but originally designated a square or street near or alongside gardens in suburban London from the early 19th century. Examples are *Cleveland Gardens*, *Gloucester Gardens*, *Leinster Gardens* and *Sussex Gardens*, all in west London. Where the term is in the singular, a specific garden is usually referred to, as in London's *Covent Garden* and *Hatton Garden*.

gate The frequently found term has two quite distinct origins. In the south of England, it usually has the normal everyday meaning, representing Old English *geat*. In the east and north of England, however, it often means 'street', 'road', and derives from Old Norse *gata*, 'path'. Thus London's *Gloucester Gate* and *Moorgate* have the regular sense of 'gate': Gloucester Gate led into Regent's Park, and Moorgate led out of the City of London to the moors. *Briggate* in Leeds, however, means 'Bridge Street' and *Micklegate* in York means 'Great Street'. The Norse sense of the name is retained in towns that were in the Danelaw, and is found as far south as Peterborough (Cam), with its *Cowgate*, *Priestgate* and *Westgate*, among others. A distinctive feature of Scandinavian *gate* names is that the specific (defining word) and the generic term frequently combine as a single word.

grove Properly a small wood or plantation, *grove* occurs in the names of some suburban streets, especially those lined with trees, or at any rate named for a particular kind of tree. Sometimes the name is simply used for variety, as for *Alma Grove*, *Bacon Grove*, *Surrey Grove* and the like, all in southeast London. In other instances there may be a reference to a historic grove, as for the better-known *Lisson Grove*, London, where the grove was in the manor of Lilestone. Examples outside London include *Flint Grove*, Bracknell (Ber), *Holly Grove*, Bromsgrove (H&W), *Linden Grove*, Hartlepool (Cle) and *La Plata Grove*, Brentwood (Ess).

hill The term is usually found for a street that ascends or descends, or that is in an elevated locality. The 'hill' itself may today not be readily apparent, however, as is famously the case with Cambridge's *Market Hill* and *Peas Hill* (see Chapter 7, p. 88). On the other hand London's *Ludgate Hill* still has a perceptible gradient from Fleet Street up to St Paul's Cathedral, and many obvious 'hills' exist elsewhere, such as *Denmark Hill* and *Herne Hill* (the one leading into the other) in south London, *St Michael's Hill*, Bristol, *Guildhall Hill*, Norwich, and the like. In several cases the street has adopted the name of an existing hill, in essence a field name.

lane The prime distinguishing feature of a *lane* is traditionally its narrowness. In the country, a *lane* is thus a narrow road between fences or hedges; in the town it runs between walls or buildings. Former narrow thoroughfares of this kind may well have been widened in later times, like London's famous *Park Lane*, which was originally a narrow way running along the eastern end of Hyde Park. But traces of the original narrowness may still be evident in current streets of the name, such as *Fetter Lane*, *Mincing Lane* and *Seething Lane*, all in the City of London.

lawn The term arose as a form of *laund* (directly related to *land*) and originally denoted an area of untilled ground. It then came to be used of a grassy area, originally one in or near a wood. In its modern use, the name in many cases deliberately evokes the regularly mown garden lawn. Its application may be purely arbitrary, as for *grove* or *mead*. Barnstaple (Dev) has several streets named with the word, including *Magdalen Lawn*, *Pathfield Lawn*, *Pilton Lawn*, *Raleigh Lawn*, *Vicarage Lawn* and *Victoria Lawn*. Some lawn names undoubtedly originated as field names.

link The term is sometimes found for a link road, that is, a shortish road connecting two others or providing a link with a commercial or other centre. One example is *Alencon Link*, Basingstoke (Ham), which joins sections of Churchill Way to provide a link with both the railway station and main shopping centre. (See this name also in Chapter 12, p. 173.) Basildon (Ess) similarly has its *Laindon Link*, connecting the town centre with the western district of Laindon. In some instances, however, *link* may have originally been a local dialect word for a ridge or bank (the source of modern golf *links*).

mead The term is an archaic word for *meadow*, and has been exploited in modern times to serve as a street name in the vicinity of an existing or former field, or in some cases simply to suggest a rustic setting. The name is characteristic of suburban residential areas. Examples are *Kilnmead* and *Smalls Mead*, Crawley (WSx), *Kingsmead*, Cirencester (Glo), *Millmead Terrace*, Guildford (Sur) and *Gorings Mead*, Horsham (WSx). Many such *meads* preserve historic field names.

mews The term is typically used of a short cul-de-sac in a fashionable area of a town or city, originally one lined by houses or apartments that have been converted from former stables. The historic reference is to *The Mews*, the royal stables at Charing Cross, London, in medieval times. These were so called as they had been built on the site of hawks' mews, a *mew* being originally a place of confinement for hawks when they were moulting (from Old French *mue*, from *muer*, 'to moult'). Many mews are still in London today, and include *Charlotte Mews*, *Chenies Mews*, *Gower Mews* and *Tottenham Mews*, all named after nearby streets. Examples outside London are *Arley Mews*,

Leamington (War), *Royal Mews* and *Nelson Mews*, Southend-on-Sea (Ess), and *Sydney Mews*, Bath (Avn).

mount The name is normally applied to a steeply rising road. *The Mount*, Guildford (Sur) is a good example. The road was formerly the main ascent to the Hog's Back. Others include *Healey Mount*, Burnley (Lan) and *Torridge Mount*, Bideford (Dev).

parade From the 18th century the name came to apply to a row of shops in a town, and so to the street where they were situated. Well-known examples are *Grand Parade*, Bath (Avn), *King's Parade*, Cambridge, *Market Parade* and *Princess Parade*, Bury (GtM), *Royal Parade*, Plymouth (Dev) and *St Augustine's Parade*, Bristol. Southern seaside resorts frequently have a *Marine Parade* running along the front, as at Brighton (ESx), Clacton-on-Sea (Ess), Eastbourne (ESx), Hastings (ESx), Southend-on-Sea (Ess), Weston-super-Mare (Avn) and Worthing (WSx).

passage The term usually applies to a narrow footpath or pedestrian way, leading either to particular houses or shops or serving as a short cut between two streets, often past a church. Many old towns have *passages*, such as *All Saints Passage*, Cambridge, *St Helen's Passage*, Oxford, *St Mary's Passage*, Stamford (Lin) and *Union Passage*, Bath (Avn).

pavement In medieval times the term basically meant 'paved way', but today by implication often denotes a throughway for pedestrian use only. Examples of the older use of the word are *High Pavement*, Nottingham, dating from the 14th century, and *Milton Pavement* and *Princes Pavement*, both in Birkenhead (Mer).

place The term normally applies to houses grouped round two or more sides of a square, often with a church on one of the sides. As such, a *place* usually lies just off an important street. An obvious example is the common *Market Place*. Where there is a church, the place will usually be named after it, as *All Saints Place*, *St John's Place*. In a few instances, often as the result of urban expansion, a place is a regular street, such as London's

Grosvenor Place, running from Grosvenor Gardens to Hyde Park Corner, or *Cambridge Place*, running off Hills Road, Cambridge. In some cases the term applies simply to a short residential cul-de-sac.

promenade The name is most familiar at seaside resorts, where it denotes a road or way favoured for walking or strolling, usually along or parallel to the coast. Familiar examples are the *Promenade* at Blackpool (Lan), Bognor Regis (WSx), Bridlington (Hum), Scarborough (NYo), Southport (Mer) and South Shields (T&W). Bournemouth (Dor) has *Durley Promenade, East Cliff Promenade, West Cliff Promenade* and *West Promenade*, among others. The term gained favour in the early 19th century, when seaside resorts were fast developing in England. Streets of this kind were familiar from French resorts, such as the famous *Promenade des Anglais* at Nice. *The Promenade* at Cheltenham (Glo) is a noted inland street of the name, and in Northampton the street skirting the southern side of the racecourse is also so called.

quay As it implies, the name denotes a street by a former or still operating wharf, usually in a coastal town or estuary port. Examples include *Beacon Quay*, Torquay (Dev), *Broad Quay*, Bristol, *Lower Quay*, Fareham (Ham) (by Fareham Lake), *South Quay*, King's Lynn (Nor) and *Town Quay*, Southampton (Ham).

rise The term usually describes a gentle hill or slope. Examples are *Boythorpe Rise*, Chesterfield (Der), *College Rise*, Maidenhead (Ber), *Crescent Rise*, Luton (Bed), *Elizabeth Rise*, Banbury (Oxf), *Orchard Rise*, Kingston upon Thames (GtL) and *Preston Rise*, Nottingham.

row The term typically relates to a narrow street with terraced houses along one or both sides, the row originally being the actual line of cottages or houses. Examples are *Chapel Row*, Bath (Avn), *Prebend Row*, Darlington (Dur), *Priory Row*, Coventry (WMd) and *Prospect Row*, Dudley (WMd). Three of these four names have religious associations, as does that of the commonly found *Church Row*.

side The term usually designates a street that runs (or ran) alongside some larger object or location, such as a park or river. London's well-known *Cheapside* ran along one side of an extensive market. The term usually blends with its descriptive word, as in the following examples: *Bankside*, Worcester (along the Worcester & Birmingham Canal), *Brookside*, Hereford, *Deneside*, Great Yarmouth (Nor), *Parkside*, Cambridge, *Moorside*, Colchester (Ess), *Mountside*, Guildford (Sur), *Riverside*, Cambridge, Chelmsford (Ess) and elsewhere, *Waterside*, Durham (by the River Wear), and *Woodside*, Harrogate (NYo).

square The term is chiefly found for the central area of a town or region, typically a rectangular market place. The latter is named *Market Square* (rather than *Market Place*) in several towns, including Aylesbury (Buc), Bishop's Stortford (Her), Corby (Ntp), Daventry (Ntp), Hanley (Sta), Harlow (Ess) and Nelson (Lan), among others. In other towns the centre is simply *The Square*.

terrace The term originally applied to a row of houses standing at a higher level above the roadway than others in a district. Later, the name came to be used for any row of houses, whether raised or not. Depending on the location and neighbourhood, *terrace* can denote either a line of simple dwellings or a fashionable row of well-appointed houses. In either case, the houses are usually contiguous (hence *terraced houses*). London's *Adelphi Terrace* (originally *Royal Terrace*) is one of the earliest examples of the name, dating from 1768.

vale A street with such a name is usually low-lying, perhaps by a stream or river, or by contrast with a nearby hill. Examples are *Brook Vale*, Chesterfield (Der), *Heath Vale*, Andover (Ham), *Horsell Vale*, Woking (Sur), *Maida Vale*, London and *Rural Vale*, Gravesend (Ken).

view The name implies a line of houses in a street that overlook a site or prominent building such as a park, hill, church or some other focal point. Examples are *Church View*, Doncaster (SYo), near St George's Church, *Hill View*, Yeovil (Som), overlooking Wyndham Hill, *Shephall View*, Stevenage (Her), facing the fields of this name, and *Station View*, Chester, overlooking the railway station.

villas The term *villa* was adopted in the 18th century to apply to a small house, detached or semi-detached, that was regarded as being better appointed than the standard town house of the day. From this, *villas* came to designate any newly built row of terraced houses, even though the houses themselves may hardly have been bigger or better than average. The classical origin of the word (Latin *villa*, 'country house') gave the name a desirable social cachet. Most towns have runs of 19th-century or later *villas* today. Examples include *Albion Villas*, Folkestone (Ken), *Clifton Villas*, northwest London, *Powis Villas*, Brighton (ESx), *Shakespeare Villas*, Nottingham and *Thornbank Villas*, Bath (Avn).

walk The term implies a street or footpath that is attractive enough for a stroll or 'promenade', without being unduly pretentious. The name is favoured by the developers of new towns and estates, so that Harlow New Town (Ess), for example, has its *Broad Walk*, *East Walk*, *Little Walk* and *West Walk*. Examples elsewhere include *Burrell's Walk*, Cambridge, *Lime Walk*, Bracknell (Ber), *Lovers Walk*, Aylesbury (Buc), *Manor Walk*, Aldershot (Ham), *North Walk*, Barnstaple (Dev) and *Rope Walk*, Bedford. These last two are in fact major thoroughfares.

wall The name is sometimes found for a street running along a castle wall, whether or not the castle still exists. The term is often found in the plural, as for *English Walls* and *Welsh Walls*, Oswestry (Shr), which respectively follow the course of the eastern and western walls of the former castle in this historic Welsh border town.

way The name often designates an important thoroughfare, especially a modern ring road or bypass circumventing the centre of a large town or city. Equally, the term can be used in the name of almost any suburban street, although the implication of a through route is still usually present. Modern examples of throughways of the name include *Churchill Way*, Basingstoke (Ham), running across the northern section of the city, *St Peter's Way*, Bolton (GtM), east of the town centre, and *Wessex Way*, Bournemouth (Dor), as a stretch of the A338 crossing the northern part of the town. Other instances of the name include *Bedingfield Way*, Bury St Edmunds (Suf), *Bideford Way*, Cannock (Sta), *Pilgrims Way*, Canterbury (Ken) (exploiting the historic name), and *Willow Way*, Christchurch (Dor). The term frequently combines with a defining word, as for the commonly found *Broadway*, *Parkway*, *Queensway* and *Ringway*. *Portway* and *Ridgeway* are names of ancient trackways adopted in some places for modern streets. In Bath (Avn) *Avon Walkway* is a long riverside walk. In touristic usage the word often designates a long-distance footpath or 'national trail'; for examples, see Chapter 2, p. 30.

In some residential districts the name is spelt *Waye*, as if to suggest a historic link. Examples are *Berkeley Waye*, *Somerset Waye*, *Blossom Waye* and *Meadow Waye*, four parallel streets in Hounslow, west London, and *Dorset Waye* and *Devon Waye*, Hillingdon, west London, two short streets opposite each another off The Rise, by Hillingdon Golf Course. (The spelling was genuine in Shakespeare's day, but has hardly been found since except in artificial revivals of this type.)

yard The term has its normal sense to denote a small enclosure surrounded by buildings, often originally warehouses or commercial premises. Examples are *Penney Yard* and *Salt Yard*, Darlington (Dur), *Sumpter Yard*, St Albans (Her) and *Sweetmans Yard*, Hereford.

There are also less common generic terms of dialectal origin, and as such normally restricted to a particular region of the country. Many of them have an identical sense, typically 'narrow lane'. They include the following:

back A term used for a small street behind buildings, often named for a larger street adjacent to it. Streets named *Back* are common in Manchester and other northern towns. For examples see Chapter 3, p. 37.

chare A term for a narrow winding lane in northern towns, especially Newcastle (T&W), so essentially corresponding to *wynd* (see below). The word probably derives from Old English *cerr*, 'turn', 'bend'. Examples include *Broad Chare*, *Manor Chare*, *Pudding Chare*, *Trinity Chare*, all in Newcastle, also *St Marys Chare*, Hexham (Ntb).

drift The word is used in some Midlands and northern towns for a lane or road along which horses or cattle were originally *driven*, that is, as a *drove* road. The name is thus a first cousin of the more generally found *drive*, above. *Drove* is itself also found (see below). Examples are *First Drift* and *Second Drift*, Wothorpe, near Stamford (Lin) (a town where *Drift Road* and the contiguous *Drift Avenue* and *Drift Gardens* are also found). Easton on the Hill (Ntp), south of Stamford, has *Ketton Drift*, a track leading to the village of Ketton (Lei).

drove An alternative term for *drift* (see above), found mainly in Fenland towns. Examples are *Meadow Drove* and *Mill Drove*, Bourne (Lin), *Conquest Drove*, *King's Delph Drove*, *New Meadow Drove* and *Straight Drove*, Farcet, and *Broad Drove*, *Great Drove*, *Hog Fen Drove*, *Leading Drove*, *Mere Drove* and *Quickset Drove*, Yaxley, the latter two places both near (and now part of) Peterborough (Cam).

rigg The word is a northern form of *ridge*, either as a dialect form of Old English *hrycg* or evolving direct from the equivalent Old Norse *hryggr*. The term is used for a street or road on an elevated site. Examples include *Castle Riggs*, Kendal (Cum) and *Four Riggs*, Darlington (Dur).

twitten A term for a narrow lane, usually a straight one between walls or hedges. The word may be related to Old English *twicen*, 'place where two roads meet', while also being an alternative form of *twitchel*, a word having the same sense.

Twitten is found mainly in Sussex, such as *Church Twitten* in Lewes (ESx). However, *The Twitten* is also a short street in Farcet, near Peterborough (Cam).

wynd The term is found in some northern towns (and also in Scotland) for a narrow winding lane or alley. Examples are *Post House Wynd*, Darlington (Dur) and, farther south, *West Bank Wynd*, Mansfield (Not). *Wiend*, Wigan (GtM) is an alternative form of the word.

Less common generic terms of this type are frequently found as simplex names preceded by the definite article, such as *The Inhedge*, Dudley (WMd), *The Slade*, Daventry (Ntp) or *The Wylde*, Bury (GtM). As indicated in the Introduction, they have also been favoured by street planners and namers in new towns, so that Basildon (Ess) has *The Fremwells*, *The Gore*, *The Knares* and *The Knowle*, among others. Many of these are dialect terms, and have been adopted from field names. *The Slade* thus denotes land in a marshy valley, and *The Gore* a triangular piece of land. The adoption of such terms from rural features is frequent. For further examples, see the Appendix, p. 206.

2. ROMAN ROADS AND ANCIENT WAYS

England is criss-crossed by a number of old roads and routes, several of them of considerable length and more than a few running in a direct, even undeviating, course for many miles. They are broadly of two types: ancient trackways, along which men and animals have travelled for thousands of years, and Roman roads, constructed and maintained by the Romans during their occupation of Britain from the 1st century BC. In some cases the Romans laid out their routes along existing trackways. In most instances, however, they constructed entirely new routes, designed to connect the towns and encampments that they had established in different parts of the country.

In modern times parts of these historic routes have been incorporated in major trunk roads and motorways. A section of Stane Street, for example, now serves as the A29 in West Sussex, while the A46 follows almost entirely the course of the old Fosse Way between Lincoln and Leicester. More extensively, Watling Street is still largely represented by the A2 and A5 of today.

The exploitation of such routes in modern times has helped to promote their names, so that *Ermine Street*, *Watling Street*, the *Fosse Way* and *Icknield Way* are now nationally familiar.

Of these, the first three are Roman roads, while the fourth is an ancient trackway. As mentioned in Chapter 1, the names of Roman roads are themselves not of Roman origin, but were mostly given by the Anglo-Saxons.

Ermine Street, which ran from Pevensey (ESx) to York, via London and Lincoln, takes its name from the *Earningas*, the Anglo-Saxon tribe whose own name means 'Earna's people'. The same tribal name lies behind that of the village of Arrington (Cam), west of Cambridge, which lies on Ermine Street (now the A14 there).

Watling Street, which ran from the east coast of Kent through the southeast and centre of England to Wales, derives its name from the *Wæclingas*, 'Wæcel's people', who gave their name to *Wæclingaceaster* ('Watlingchester'), today more familiar as St Albans (Her). (The city was renamed for the 3rd-century saint and martyr St Alban, who is buried there.)

The *Fosse Way*, as already mentioned, takes its name from the ditch (Latin *fossa*) that ran on both sides of it. The road itself followed a route from near Sidmouth (Dev) to the Humber, via Ilchester (Som), Leicester and Lincoln. North of Lincoln, its course coincides with that of Ermine Street. It has been suggested that the ditches may have been intended for defensive purposes. Early documents refer to the road simply as *Fosse*, and *Fossewey* is not recorded until the mid-15th century.

The *Icknield Way* originally ran from the Wash to the south coast, and today can be traced from Thetford (Nor) to Baldock (Her), and thence below the northern escarpment of the Chilterns to Streatley (Ber), and on westwards along the Ridgeway to Avebury (Wil). In modern terms the first part of this route is mainly that of the A11 and A505. The name of the trackway is not Anglo-Saxon, but pre-English, and of uncertain origin. It has been popularly linked with the *Iceni*, the British tribe whose capital was at *Venta Icenorum* (modern Caistor St Edmund (Nor)), in an area where the Icknield Way begins. Even so, any etymological link between the folk name and that of the road remains unfounded. The name also lies behind that of *Icen Way*, a modern street in Dorchester (Dor). It was given from the fact that the name *Icknield Way* was adopted for the old Roman road (now the A35) that ran through Dorchester from London to Exeter. The spelling of the modern name appears to have been based on the spurious Latin form *Via Iceniana*, introduced by antiquarians in the 18th century.

Another well-known Roman road is *Akeman Street*, which branched off from the Fosse Way at Cirencester (Glo) and then ran east to Tring (Her), passing south of Bicester (Oxf). In Anglo-Saxon times, however, *Akeman Street* was also used for what is now the Fosse Way as the road between Cirencester and Bath (Avn), and the name must certainly be linked with the latter city, which in the 10th century was recorded as *Acemannes ceastre* ('Akemanchester'). This suggests that the whole of the Roman road from Tring to Bath was known at one time as *Akeman Street.* The name itself means 'Aceman's street', from a personal name, although it is not known who Aceman was. Because Bath gained early fame for its hot springs, used in treating rheumatic diseases, the name popularly became associated with *ache*, as if Bath itself was the 'Roman town of aching men'.

These names had become so familiar generally by medieval times that they were adopted for Roman roads and old trackways elsewhere in the country. *Ermine Street*, for example, is also the name of the Roman road that ran from Gloucester to Silchester (Ham), the first section of which is the modern A417 and A419 through Cirencester (Glo) and Swindon (Wil). *Watling Street* is now also the name of the Roman road from Chester to Monmouth, via Wroxeter (Shr), as well as that of the Roman road from York to Corbridge (Ntb). The *Icknield Way* gave the name of the Roman road known as *Ryknield Street*, originally said to have run from St Davids, in Wales, to York, but now regarded as a branch of the Fosse Way extending from Bourton-on-the-Water (Glo) to a point near Rotherham (SYo). (The *R*- of the street name apparently evolved from an interpretation of Middle English *at there Ikenilde strete* as *at the Rikenilde strete*, in other words the final *-r* of the feminine dative case ending of the definite article was taken to be the first letter of the following syllable.) It was also adopted, as mentioned, for the Roman road from London to Exeter (Dev).

Akeman Street, in the west of England, has in turn had its name transferred to a Roman road on the other side of the country, in East Anglia, where it runs between Arrington on Ermine Street to Denver, near Downham Market (Nor), via Cambridge and Ely (Cam) (today's A603 and A10).

Roman roads are also indirectly indicated by the names of places that stand on or near them. Almost all modern towns and villages with names based on Old English *strǣt*, modern 'street', imply the presence of a local Roman road. Examples include the many places called *Stratton* and *Stretton* (*Stratton-on-the-Fosse* (Som) being on the Fosse Way), the numerous towns and villages called *Stratford* (that in east London is on the Roman road to Colchester (Ess)), and such places as *Chester-le-Street* (Dur), *Streatham*, southwest London, *Streatley*, near Pangbourne (Ber) (by the Icknield Way, as mentioned), and *Sturton*, near Brigg (Hum) (off Ermine Street).

Some ancient trackways likewise appear with the same name in different parts of the country. One of the most common is *Portway* (or *Port Way*), whose best-known example is the Roman road that runs between Silchester (Ham) and Old Sarum (Wil). Wherever it occurs the name means 'way leading to a town', from Old English *port*, 'town', later 'market town', as for New*port* Pagnell (Buc). It is usually not clear which particular town is meant, however. Other common trackway names include *Broadway*, *Holloway* and *Ridgeway* (also each spelled as two words). The meanings of such names are self-explanatory: 'broad way', 'hollow way' and 'ridge way'. This last is familiar as the *Ridge Way* which runs from the Vale of Pewsey (Wil) along the northern edge of the Wiltshire and Berkshire Downs to the River Thames at Streatley (Ber), where it joins the Icknield Way. In Cambridgeshire there are additionally a number of old narrow roads known as *Smallway*, by contrast to a neighbouring *Broadway*.

Trackways named *Saltway* are prominent in the county of Hereford and Worcester. One well-known route so called led from Droitwich (H&W) northwards to Ryknield Street south of Birmingham and southwards to Worcester, Tewkesbury (Glo) and Gloucester. As it implies, the name means 'salt way'. Droitwich and its neighbourhood were the only places in central England where salt was produced in medieval times. It was carried along *Saltway* routes by packhorse to the places where it was needed. Many of these names survive today in the names of farms: *Broadway Farm*, *Ridgeway Farm* and so on.

Stane Street ('stone street') is a Roman road that ran from London to Chichester (WSx). As for other such roads, its name was later adopted elsewhere, in this case for the Roman road from Bishop's Stortford (Her) to Marks Tey, near Colchester (Ess), passing through Dunmow (Ess) and Braintree (Ess) and today followed by the A120 (see also *Hare Street* below).

Certain individual ancient routes have names that are at least locally familiar. The prehistoric *Harrow Way* (or *Harroway*) runs across southern England from east to west, extending from Kent to Wiltshire (where its terminating point may have been Stonehenge) and crossing the Thames at Streatley (Ber). It is said to be Britain's oldest road, and its name accordingly means simply 'old road', from Old English *hār*, 'old', 'grey' (the source of modern *hoary*). The road that is now *London Road*, Newbury (Ber) (the modern A4, running eastwards from the town centre), is recorded in a document of 1550 as 'the King's high waye, called Harrow Waye, leading from Newbury to Reddinge'. This was not the actual course of the Harrow Way, which ran to the south of the present town, but the name was undoubtedly transferred from it.

A similar ancient track is the *Lunway*, running east and west across central Hampshire between the Harrow Way to the north and the south Hampshire Ridgeway to the south. Its name, probably meaning 'London way', is preserved in that of the *Lunways* Inn, Micheldever, north of Winchester (Ham). (If the road were further north it would be tempting to derive its name from Scandinavian *lúndr*, 'wood', 'grove', so that it would be a 'woodland way'. A now lost *Lunway* in Cambridgeshire certainly seems to have had this sense.) In East Anglia *Peddars Way* is an old route, still traceable in places, that runs from Ixworth, north of Bury St Edmunds (Suf), to the Wash at Hunstanton (Nor). Its name derives from *pedder*, a dialect (and now mainly Scottish) word with the same sense as *pedlar*, to which it is presumably related. The *Sugar Way* runs for some 13 miles from the North Berkshire Downs to the Marlborough Downs (Wil). Its name seems to imply the equivalent of a *Saltway*, but a 16th-century record of the name as *Scocera Weg* suggests that its true origin is in Old English *scēacere*, 'robber' (compare modern German *Schächer* in the same sense), so that it was the 'robbers' way'. The same source gave the name of *Sugar Hill* near Aldbourne (Wil), which lies near the Sugar Way.

The *Wansdyke* is essentially an ancient earthwork rather than a trackway. It originally ran from Berkshire across Wiltshire to Portishead (Avn), west of Bristol, and was constructed for purposes of defence. Its name means 'Woden's dyke', implying that the famous Anglo-Saxon god was either believed to have constructed the dyke or that he was regarded as its tutelary deity. Much of the dyke has now been obliterated, but sections of it are well preserved and serve as a footpath, notably in north Wiltshire.

A similar name to that of the Wansdyke in Wiltshire is that of *Grim's Ditch*, a prehistoric earthwork that runs through Grovely Wood northwest of Wilton (Wil) and that gave the name of the village of *Ditchampton* here. The name is found elsewhere for an ancient earthwork, and is basically the same as that of the Wansdyke itself, since *Grim*, from Old English *grīma*, 'mask', is another name for Woden.

Also in Wiltshire, at least as a local name, is *Green Street*, a seven-mile stretch of an ancient trackway along the Marlborough Downs between Avebury and Marlborough. This is an Anglo-Saxon *herepæth*, or military road (from *here*, 'army' and *pæth*, 'path', 'track'), used for the movement of troops and by border patrols. An alternative name for this particular trackway actually is *Herepath*, and another such military road further south in Wiltshire, in the Vale of Pewsey, has its name preserved in that of *Harepath Farm*, east of Bishops Cannings.

The *Via Devana* is a spurious Latin-sounding name for the Roman road that was said to run from Colchester (Ess) to Chester, the latter having the (genuine) Roman name of *Deva*. The name was actually devised in the 18th century for the road by Dr Mason, Woodwardian Professor of Geology at Cambridge University. A stretch of the *Via Devana* familiar to motorists is the A504 road from Cambridge to Godmanchester, south of Huntingdon (Cam), where it meets up with Ermine Street (the modern A14).

The *Via Gellia*, despite its Latin-style name, is not even a Roman road, but the road (now a section of the A5012) that was constructed in the late 18th century along a wooded ravine between Cromford and Grangemill, near Wirksworth (Der). The builder of the road was Philip *Gell* of Hopton Hall, who named it after himself. (When the textile firm of William Hollins & Co established a cloth mill on the Via Gellia in the 19th century, the road name was adopted for the famous brand of cloth known as *Viyella*, although in fact this particular cloth was not produced at the Via Gellia Mills.)

The names of other trackways and sections of Roman road are generally less well known, but are still of interest. *Ashwell Street* is an ancient Roman-style trackway that runs from Ashwell (Cam), near the border with Bedfordshire, to Melbourn (Cam), crossing Ermine Street on its course. It takes its name from the former village. *Bathamgate* is a Roman road that runs from Brough (Der) to Buxton (Der). Its name means 'road to the baths', referring to the natural springs at Buxton, which was known to the Romans as *Aquae Arnemetiae.*

Buckle Street is the name of a section of Ryknield Street which runs from Bidford, west of Stratford-on-Avon (War), to Aston Subedge, north of Chipping Campden (Glo). Its name derives from the Anglo-Saxon female personal name *Burghild*, doubtless referring to a woman who owned land somewhere along its route. *Car Dike* is a Roman road that runs south from Sleaford (Lin) into Cambridgeshire, passing through Bourne (Lin) on its route. It derives its name from the Scandinavian personal name *Kárr*, probably referring to a Viking settler who had some connection with this ancient dyke, which is believed to have formed part of a Roman canal system from Lincoln to Cambridge.

More romantic is the *Devil's Highway* (Ber), a section of the Roman road that ran from London to Silchester (Ham), and that passes through Easthampstead, now a district of Bracknell (Ber).

Its name is recorded no earlier than the 18th century. Similar to this is *Dhoul's Pavement*, a section of the Roman road that runs along Blackstone Edge east of Littleborough (GtM). Its name means 'the devil's pavement', from a dialect form of Old English *deofol*, 'the devil'. These names should be compared to those of *Wansdyke* and *Grim's Ditch* (above, p. 22), which also relate to fearsome deities.

The *Doctor's Gate* is a Roman road in Derbyshire that runs from Brough to Melandra Castle, near Glossop. The name is recorded as *Docto Talbotes gate* (*sic*) in 1627, although the identity of the particular Dr Talbot who apparently gave his name to the road remains uncertain. *Gate* here has the sense 'road'. In Gloucestershire, *Hare Lane* (formerly *Hare Street*) was a name used for the road from Worcester to Gloucester (the present A38). It derives from Old English *here*, 'army', so that the road was in essence a military highway. Today *Hare Lane* is the name of one of the streets in Gloucester itself, running parallel with Worcester Street, north of the city centre. In Hertfordshire, *Hare Street*, of the same origin, is an alternative name for the Roman road known more generally as *Stane Street* ('stone street'), running between Bishop's Stortford and Baldock.

In Warwickshire, *Haydon Way* is the name of a stretch of Ryknield Street between Studley and Alcester, south of Redditch (H&W). It was recorded as *Havedway* in the late 13th century, and presumably derives from Old English *hēafod*, 'head', so that it was the 'chief way'. *Hereward's Street* is an old road that seems to have run from Chesterfield (Der) to Ashbourne (Der), but which now can be traced only in sections near the village of Ashover, west of Clay Cross. The name means what it says, 'Hereward's street', although it is not known who Hereward was.

The road that is now the A31 running west from Guildford (Sur) is familiarly known by the name of the ridge along which it lies, the *Hog's Back*. The reference is to the appearance of the chalk ridge as seen from the lower land on either side. The name is a modern one, first recorded no earlier than 1823. The earlier name of the ridge was *Guildown*, recorded in the late 12th century as *Geldedon* and revived in modern times for *Guildown Road*, extending westwards towards the Hog's Back from the Portsmouth road, south of the city centre. *Guildown* probably took its name from Guildford itself; that is, it was really *Guildford Down*.

A road name of more definite date than most is *Ireton's Way*. It is that of the road built by General Ireton in the Civil War to convey troops from Chatteris (Cam) to Ely (Cam) (the present A142), and is first recorded in 1654. *Leeming Lane*, in Nottinghamshire, is an old road running from Mansfield (Not) to Worksop (Not). The name seems to have been adopted from that of *Leeming Lane* in North Yorkshire, where it is used of the Roman road that ran north from Boroughbridge (NYo) to Catterick (NYo), along the course followed by the present A1. Its own name comes from the village of *Leeming* on its route, northeast of Bedale (NYo).

Maiden Way is a Roman road running from Kirkby Thore, northwest of Appleby (Cum) to Alston (Cum). It almost certainly takes its name from the former *Maiden* Castle in Kirkby Thore itself, on the site of the present Low Abbey. The castle's own name, like those of castles elsewhere in the country, including the famous one in Dorset, indicates that it was 'virginal', that is, it had never been taken or breached.

The *Mareway*, in Cambridgeshire, forms a parish boundary for some 10 miles from Ermine Street to Red Cross, southeast of Cambridge. Its name refers to its delimiting function, and means 'boundary way', from Old English *gemǣre*, 'boundary' (modern but now archaic English *mere*). The same name occurs elsewhere, for example *Mereway Road*, Twickenham, southwest London, which also runs close to a parish boundary.

Regia Via is a name found in various parts of the country for a locally important old road. Warwickshire has at least three such roads, one running northeast to the lost village of Hopsford between Shilton and Withybrook, another, now a cart track, linking Radford Semele, east of Leamington, with neighbouring Whitnash, south of the town, and a third as the Stratford-on-Avon to Henley-in-Arden road (the present A34) through Wootton Wawen. The name is Medieval Latin for what is literally 'royal road', but the sense corresponds to the modern *high road* or *High Street*, ie it denotes a main road. There may have been some subsequent confusion with *Ridgeway*, and indeed this was the name for the old road that ran from Whitnash, just mentioned, to the nearby Fosse Way.

In North Yorkshire, the Roman road that ran north from Toulston, southwest of York, to a point on the York to Boroughbridge road (the present B6265) near Whixley came to be called *Rudgate*. The name, found in similar forms elsewhere in Yorkshire, probably means 'Rudda's road', from an Anglo-Saxon personal name. Further west, the old road that ran between Lambrigg and Firbank, along the west bank of the River Lune in Cumbria, gained the local name of *Scotch Lane*, apparently since it was the route along which Scotch cattle were driven. However, its medieval name was *Galwaithegate*, 'road to Galloway', showing that a link with Scotland itself cannot be ruled out. The nearby M6 is today the main road to Scotland here, east of Kendal (Cum).

Wade's Causeway, also in North Yorkshire, is perhaps Britain's best surviving Roman roadway. It is a stretch of just over a mile in length running across Wheeldale Moor north of Pickering, and was at one time part of a 30-mile route from Malton to Mulgrave Castle, northwest of Whitby. It takes its name from a famous hero of antiquity, mentioned by Chaucer, who is said to have been the son of Wayland the Smith and to have had a magic boat called Wingelock. The story connecting Wade with the road is that he and his wife set out to construct a shorter route to the moors for the benefit of the latter when she went milking. The wife dropped her load of stones while building the road, and the Causeway was the result.

Similar to Scotch Lane is *Welsh Way*, a Roman road that ran northwest from Fairford (Glo) towards Wales, crossing

Akeman Street. The name is self-explanatory, although it has been suggested that more specifically it was a route to London used by Welsh cattle drovers in order to avoid Cirencester (Glo). It was earlier known as *Street Way*, testifying to its Roman origin. Also in Gloucestershire is the *White Way*, a ridgeway running from Cirencester towards Compton Abdale, east of Cheltenham. The name may refer to the fact that it was a branch of a *Saltway*.

A section of the Via Devana in Cambridgeshire, southeast of Cambridge, where it crosses the Icknield Way, is known as *Wool Street*. Medieval records of the name, such as *Woluestrate* in 1334, suggest that the original meaning was 'Wulf's street', from an Anglo-Saxon personal name, although it is not impossible that the sense was 'wolves' street'. Wolves could have been numerous here at one time.

It is hardly surprising that many ancient roads and trackways are called simply *Old Street* or (rather more precisely) *Roman Road*, and many names of this type in towns today could relate to a genuinely historic route. London's *Old Street*, running through Finsbury and Islington, is believed to be part of a Roman road that ran through Bethnal Green to the River Lea, where it is represented by *Roman Road* itself. Similarly *Old Street*, running north from Chieveley, north of Newbury (Ber), is probably an ancient way, while *Roman Way* is a section of the Roman road (now the A656) that runs north to the east of Leeds in West Yorkshire. At Stamford (Lin), *Roman Bank* represents a short section of Ermine Street at the point where the Roman road crossed the River Welland on its route to Lincoln.

Some stretches of former Roman roads are known simply as *The Street*, preserving the Latin origin of the name. One example is the Derbyshire road, now the A515, that ran southeast from Buxton to a point west of Wirksworth. The name is no longer in current use, but dates from at least the early 15th century. The road's starting point can be regarded, however, as Buxton's *High Street*, to the south of the town centre, and The Street was itself recorded as the *High Streete* in the 17th century.

Mention should also be made of two of the best-known long-distance English roads, the *Great North Road* and the *Great West Road*.

The *Great North Road* is the main route from London to Scotland up the eastern half of the country, today followed largely but not exclusively by the A1. It originally led out of London by the present A1000 through Barnet (GtL), Potters Bar (Her), Hatfield (Her) and Welwyn (Her), and then continued north by what is now the B197 (alongside the M1) via Stevenage (Her) and Baldock (Her). In the latter town it crossed the Icknield Way. North of Biggleswade (Bed) it followed the course today mainly represented by the A1 as far as Markham Moor (Not), a lengthy stretch coinciding with Ermine Street between Huntingdon (Cam) and Grantham (Lin). From Markham Moor it followed the route of today's A638 to Doncaster (SYo), after which it is more or less represented by the A1, although the modern road bypasses many of the towns, such as Doncaster itself, Wetherby (WYo) and Boroughbridge (NYo), that the Great North Road passed through. North of Boroughbridge it ran along the course of today's A167 as far as Darlington (Dur) and Durham, then what is now the A6127 to Newcastle (T&W). The last section of the Great North Road is again approximately represented by the modern A1, passing through Morpeth (Ntb) and Alnwick (Ntb) to Berwick-upon-Tweed (Ntb).

The name itself is relatively modern, with one of its earliest records in a tithe award of only 1849. Today it is in use for various streets in towns and cities along its former route, such as the built-up A1000 road between Barnet (GtL) and Hatfield (Her), the B197 south of Welwyn (Her), a minor street south of Stevenage (Her) town centre, and the A1 as it leaves northern Newcastle (T&W).

The *Great West Road* does not really belong in this chapter, as it is neither a Roman road nor an ancient way. But in a sense it is the modern equivalent, so deserves inclusion. It was the name given to the road, now the first part of the A4 from London to Bath (Avn), that was built in the 1920s to provide a better and faster route to the west, avoiding Chiswick's congested High Street. It lies roughly parallel to the old Bath road, the A315 of today, and its initial stage ran from the present Chiswick Flyover to Staines, a distance of 8½ miles. Today the name applies not only to the section from Gunnersbury to Lancaster Place, Hounslow, where it is joined by Bath Road (now the A3006), but to its postwar extension eastwards to the centre of London, through Chiswick and Hammersmith. Officially, however, the eastern extension bears the name only for the stretch between the Hammersmith Flyover and the Hogarth Roundabout, Chiswick, where it is joined by the Great Chertsey Road (the A316). The name, recorded on Bacon's *Large Scale Atlas of London and Suburbs*, published in about 1927, was undoubtedly modelled on that of the *Great North Road*.

Ancient ways and Roman roads have their modern equivalents in a number of long distance walking or bridleway routes. Many of them actually follow historic trackways, and several take the course of long-established 'rights of way'.

The best known are the so-called 'national trails' created by the Countryside Commission, who establish them, waymark them (with directional arrows), maintain them, and popularise them through promotional literature. There are currently 15 such routes in England. One is the famous *Pennine Way*, the first of England's long distance footpaths (and the third longest), set up in 1965 and running north from Edale (Der) along the 'Backbone of England' to end just over the Scottish border at Kirk Yetholm (Borders). Another, much more moderate, incorporates the *Peddars Way*.

Next come the 'regional routes', around 100 in number, which are trails created by local authorities, usually county councils, with the support of the Countryside Commission, who meet approximately half the cost.

Finally there are the unofficial trails, the long distance footpath routes devised by groups of walkers, such as branches of the Ramblers Association, or even by enterprising individuals.

While many routes are well defined, especially the bridleways, in some cases there may be no form of clear-cut course to follow. The Pennine Way, for example, is undefined for lengthy stretches, while the *Two Moors Way* (see below) requires the use of a compass over parts of its course, as there is little or nothing to see on the ground by way of an actual track.

The names of many such routes are geographically descriptive, such as the *Essex Way*, from Epping Forest (Ess) across the county to Harwich (Ess), the *Nene Way*, mainly along the course of the River Nene, from Badby (Ntp) to Wansford (Cam), and the *South Downs Way* from Winchester (Ham) to Eastbourne (ESx). However, some have more interesting and original names, despite the almost standard generic word *Way*. A selection follows.

Angles Way, from Knettishall Heath (Suf), where the *Peddars Way* leaves off, to Great Yarmouth (Nor); named for the Angles who gave the name of East Anglia here (as well as that of England itself).

Centenary Way, from York to Filey (NYo); created by the North Yorkshire County Council and named to mark 100 years of its own existence.

Clarendon Way, from Winchester (Ham) to Salisbury (Wil); named for the remains of the medieval Clarendon Palace, east of Salisbury, and the classic mansion Clarendon House now there. It was this Clarendon that gave the title of the Earls of Clarendon, notably that of Edward Hyde, 1st earl of Clarendon (1609-1674), for whom in turn are named the historic Clarendon Code, the Clarendon Laboratory, Oxford, and the Clarendon Press, Oxford.

Coast to Coast Walk, from St Bees Head (Cum) to Robin Hood's Bay (NYo); named for its course right across England from the Irish Sea to the North Sea. The walk, which was devised by the best-selling writer on fell walking, Alfred

Wainwright, is one of the most popular in Britain, despite being unofficial.

Greensand Way, from Haslemere (Sur) to Limpsfield (Sur); named for the belt of greensand that it follows across Surrey between the North Downs and South Downs. (It should not be confused with the somewhat shorter *Greensand Ridge Walk* which runs from Leighton Buzzard (Bed) to Gamlingay Cinques (Cam). This similarly follows a ridge of green sandstone.)

Harcamlow Way, from Harlow (Ess) to Cambridge, passing through Saffron Walden (Ess) in a figure of eight circuit; named for *Harlow*, at its southern end, and *Cam*bridge, at its northern. The footpath walk was opened in 1980 to mark the 10th anniversary of the West Essex Group of the Ramblers Association.

Heart of England Way, from Cannock Chase (Sta) to Chipping Campden (Glo); named for its course through the central counties of the Midlands.

Hereward Way, from Peterborough (Cam) to Ely (Cam); named for the 11th-century Anglo-Saxon rebel Hereward the Wake, who plundered Peterborough and defended the Isle of Ely against William the Conqueror.

Jubilee Way, from Melton Mowbray (Lei) to Woolsthorpe (Lin); opened as a footpath through the Vale of Belvoir in 1977 and named for the Silver Jubilee of the Accession of Queen Elizabeth II that year. The path joins the Viking Way west of Grantham (Lin).

King Alfred's Way, from Portsmouth (Ham) to Oxford; named for the 9th-century king of Wessex and overlord of England, whose capital was at Winchester, through which it passes.

Leland Trail, from Stourhead (Wil) to Stoke sub Hamdon (Som); named for John Leland (?1506-1552), royal librarian to Henry VIII, who at some period between 1535 and 1543 made a journey through Somerset. His detailed account of his route has enabled the trail to follow closely in his footsteps.

London Countryway, encircling Greater London as a sort of walker's version of the M25, and named for its rural route. Its course can be picked up at any point, but if begun in a clockwise direction at Box Hill (Sur), passes through Windsor (Ber), Marlow (Buc), St Albans (Her), Epping Forest (Ess) and Gravesend (Ken).

Lyke Wake Walk, from Osmotherley (NYo) to Ravenscar (NYo); named for the title of a 15th-century Yorkshire funeral dirge, one of the most familiar in England. The name was given humorously by the local walkers who pioneered the route.

Robin Hood Way, from Nottingham to Sherwood Forest (Not); named (no doubt inevitably) for Nottingham's legendary outlaw. The route starts from his statue by Nottingham Castle and ends in Edwinstowe (Not), in whose church Robin Hood and Maid Marion are said to have been married.

Saints Way, from Padstow (Cor) to Fowey (Cor); named for the route, from the north Cornish coast to the south, said to have been followed by Christian missionaries from Ireland who came to convert the Cornish. The trail is believed to follow a prehistoric trading route used by early merchants who wished to avoid rounding Land's End in their frail craft.

Sandstone Trail, from Frodsham (Che) to Grindley Brook (Shr); named for the ridge of sandstone that it follows as it crosses Cheshire from north to south.

Saxon Shore Way, from Gravesend (Ken) to Rye (ESx); named for the coastline here, which in the 4th century AD was subject to attacks by Saxon raiders and which was governed by a Roman commander known as *Comes litoris Saxonici*, 'Count of the Saxon Shore'.

Swan's Way, from Salcey Forest (Ntp/Buc) to Goring (Oxf); named for the swans seen in the valley of the Great Ouse, which it crosses, and again at its finishing point on the Thames (where swan-upping takes place), with a punning literary allusion to Marcel Proust's *Swann's Way*. The swan is also the emblem of Buckinghamshire County Council.

Three Shires Way, from Tathall End (Buc) to Grafham Water (Cam); named for the three counties of Buckinghamshire, Bedfordshire and Cambridgeshire (or, historically, Buckinghamshire, Bedfordshire and Huntingdonshire), through which it passes.

Torpel Way, from Stamford (Lin) to Peterborough (Cam); named for the medieval manor of Torpel near Bainton (Cam), through which it passes. The manor was the seat of a Norman baron of Peterborough, and remains of his castle still exist.

Two Moors Way, from Ivybridge (Dev) to Lynmouth (Dev); named for Dartmoor, south Devon, where it starts, and Exmoor, north Devon, where it ends.

Viking Way, from the south end of the Humber Bridge (Hum) to Oakham (Lei); named for the Vikings who settled in this part of the country (establishing the Danelaw here), and who left their memorial in many place-names along the route, notably those ending in *-by*, such as South Ferriby, Barnetby, Bigby, Grasby, Normanby, Walesby, Goulceby, Scamblesby and Fulletby, all before the trail reaches Lincoln. The name is not meant to imply that the Vikings actually followed this route.

Wayfarer's Walk, from Inkpen Beacon (Ber) to Emsworth (Ham); named for the many field paths and old tracks that it follows across Hampshire, including a section of the *Portway* leading to Alresford.

Wealdway, from Gravesend (Ken) to Beachy Head (ESx); named for the Weald, the region of the former forest in Kent and Sussex between the North Downs and the South Downs.

Weavers' Way, from Great Yarmouth (Nor) to Cromer (Nor); named for the region of Norfolk where weaving once flourished as a cottage industry.

Wolds Way, from Hessle (Hum) to Filey (NYo); named for the Yorkshire Wolds, the chalk hills north of the Humber across which the route runs.

Finally, mention should be made of the *Pilgrims Way*, the path from Winchester (Ham) to Canterbury (Ken) that was the ancient trackway probably used as a route by medieval pilgrims travelling to the shrine of the 12th-century saint and archbishop Thomas à Becket in Canterbury Cathedral. The path follows the southern slope of the North Downs, crossing the River Wey just south of Guildford (Sur) at the point marked by the modern road of the same name. Today a section of the route is followed generally by the M25 between Reigate (Sur) and Oxted (Sur). The historic name is relatively recent, and is recorded no earlier than the second half of the 18th century. In Edward Hasted's *The History and Topographical Survey of the County of Kent* (1778), the route is referred to as *The Upper Pilgrim Road* near Otford, north of Sevenoaks (Ken), and as *Pilgrim Road* near Halling, southwest of Rochester (Ken).

Long distance road and way names of this type, both ancient and modern, essentially form a special category. The most familiar kinds of names are those found for town streets from about 1100. These are therefore the names that will chiefly be found in the chapters that follow.

3. SELF-DESCRIPTIVE NAMES

Many street names relate to the road or street itself, giving information about its importance, size, nature or function.

One of the most familiar of all street names is *High Street*, found in most towns as that of the present or past commercial centre, and still the expected locality of shops and businesses, from estate agents to chain stores.

This name should not be taken literally. It does not mean that the street is elevated but that it is important. *High* therefore has the same sense as in *highway* (as distinct from *byway*). Again, although a High Street may be the principal one in a town, it does not follow that it is the oldest. Most modern High Streets were originally named for the town or village in which they arose, so that High Street, Harrow, west London, for example, was *Harwestrete* in the 15th century, and the present High Street in Homerton, east London, was known as *Humberton Street* in the mid-17th century. It is true that *High Street* is occasionally found for sections of Roman roads, as for that north of Appleby (Cum), but this is the exception rather than the rule. Some High Streets arose as the street in a village where a market was regularly held, so became important in that sense. The High Street in Uxbridge, west London, was thus known as *The Markett Place* in the late 16th century. Oxford's famous High Street ('the High') is recorded under this name no earlier than the 17th century. It was previously known as *Eastgate Street*, as it led from the east gate of the city to Carfax. (It is possible, however, that a road recorded as *altus vicus* in the late 14th century could be the present High Street.)

A few High Streets date back to medieval times, even so. One of the earliest must be that in Barnet, north London, which is recorded as *Le Heyestrate* in 1348.

In the north of England the equivalent name to High Street is frequently found as *Highgate*, for example in Kendal (Cum) and Beverley (Hum). In York and Pontefract (WYo) the principal street is *Micklegate*, a name of Scandinavian origin meaning 'great street'. *King Street* can also sometimes equate with High Street, as in Chippenham (Wil), although this town also has an actual *High Street*, recorded in Medieval Latin as *alta strata* in 1326. King Street, Hammersmith, west London was similarly known earlier as *The Highway*. It was long one of the main roads from London to the West Country, before the coming of the *Great West Road* (see Chapter 2, p. 28). York's *Coney Street* is also in this category, as the name has its derivation in Old Norse *kunungr*, 'king'. It dates from as early as the 12th century, when it was recorded as *Cuningestret*.

In some towns the name for an important street may be *Main Street*, as in Bingley (WYo) and Rotherham (SYo), although the latter equally has its High Street. Elsewhere the main street may be known as *Fore Street*. This name is particularly common in the West Country, where it is found in Axminster (Dev), Bridgwater (Som), Brixham (Dev), Chard (Som), Exeter (Dev), Exmouth (Dev), Ilfracombe (Dev), Newquay (Cor) and St Ives (Cor), among other towns. Several of these also have a High Street, however. Hertford's Fore Street was *The Highe Streete* in the early 17th century, and Fore Street in Trowbridge (Wil) was probably the *alte strate* recorded (in a genitive form) in a document dated 1468. In Edmonton, north London, Fore Street was similarly *Heghestrate* in the 14th century. *Fore* properly has the sense of *before* in this name, referring to a street in front of a city wall or rampart. An illustration of this meaning is provided by Fore Street in the City of London, which ran beside and along the city wall (now represented by *London Wall* to the immediate south of it). In Exeter, too, *Fore Street* is a continuation of High Street outside the city wall.

A similar name is *Foregate*, which literally means 'street before the gate', as for *Foregate Street*, Chester. Worcester has *Foregate Street* running north out of *Foregate*, both street names having this same origin. The implication is thus not necessarily that the street is an important one, although all three streets mentioned here are central and so commercially significant.

A less important street located behind Fore Street in a town is sometimes called *Back Street.* This is so in Hertford and Trowbridge (Wil). Sherborne (Dor) has a *Back Lane* (see Chapter 8, p. 117) but its main street is *Cheap Street* ('market street'). *Hospital Road*, which runs parallel to the latter, was also formerly *Back Lane*. In the north of England, however, notably in Manchester and neighbouring towns, *Back* is prefixed to the name of a smaller street running off a larger one, with the name itself that of its 'big brother'. Thus *Back Balloon Street* is an extension of *Balloon Street*, *Back Piccadilly* is adjacent to *Piccadilly*, and so on. Manchester has dozens of such streets.

Nottingham has a High Street, but also a *High Pavement*. Contrasting with the latter is the city's *Low Pavement. Pavement* here means 'paved way', not 'footpath'. It has been suggested that London's *High Holborn* could be understood literally. But its name was recorded as *Holebrune Strate* in the late 12th century, showing that *High* relates to its importance, not its elevation. *Lowgate*, Hull (Hum) (originally *Marketgate*) was at one time the back street leading to *High Street* (formerly *Highgate* or *Hull Street*).

Streets called *Upper* and *Lower* may be named according to their importance or their topographical location. Logically, one would expect a *Lower Street* (or *Nether Street*) in a town with an *Upper Street*, although this is not always the case. *Upper Street*, Islington, northeast London at one time had a *Lower Street* to contrast with it. The latter is now named *Essex Road*, however. But in central London one can still find *Upper James Street* and *Lower James Street*, *Upper John Street* and *Lower John Street*, while *Upper Thames Street* has *Lower Thames Street* as its continuation below London Bridge.

Broadway and *Broad Street* are names that relate more to the width of a road than necessarily its importance. Hammersmith *Broadway*, west London, is one of the best known, although the name is first recorded only in 1813. Towns with a *Broad Street* include Abingdon (Oxf), Bristol, Oxford and Stamford (Lin). The well-known Oxford street ('the Broad') was formerly

Horsmongeres-streta (in the early 13th century), as it was where horses were sold. The modern name is found no earlier than the mid-18th century. In Bristol, *Broad Street* has the contrasting *Small Street* as its neighbour, small here meaning 'narrow'. Both names date from the 13th century. Further north one can find *Broadgate* as the equivalent name, as in Lincoln, which also has *The Strait*. This was originally the name of the narrow passage at the lower end of *Steep Hill* (see below), but is now the standard street that leads towards it. *The Strait* does not appear to have been deliberately so named (only in the 18th century) in contrast to *Broadgate*, although the two streets are not far apart.

In between High and Low (Higher and Lower) can come a *Mid* or *Middle*. Nottingham's High Pavement and Low Pavement are thus accompanied by *Middle Pavement*. A *Mid* name can of course relate to a street in the centre of a town, without any obvious contrasting name. This is the case in Newark (Not), which has a *Middle Gate* by its central Market Place. Peterborough (Cam) has its *Midgate* as a central street. Its earlier name was *Howegate*, 'street by the burial mound', from Old Norse *haugr*, 'hill', 'mound'.

Of all purely locational street names, *North Street*, *East Street*, *West Street* and *South Street* must be among the most common. Their northern and eastern equivalents are *Northgate*, *Eastgate*, *Westgate* and *Southgate*, although these particular names often relate to actual gates in the south of England, as for *Eastgate Street* already mentioned for Oxford. Wherever a 'compass' name of this type is found, a contrasting name will usually also exist. Some towns have all four, perhaps meeting centrally at a market cross, as in Chichester (WSx). The city developed from a Roman town, where a crossroads of four straight streets like this was standard. *North Street* and *South Street* in Guildford (Sur), on the other hand, were earlier respectively *Lower Back Side* and *Upper Back Side*. They are first recorded, however, as *the north town ditche* and *South Ditch*, and the two streets marked the northern and southern limits of the old town. Leeds (WYo) has its *Eastgate* and *Westgate*, but no

Northgate or Southgate; Pontefract, southeast of Leeds, has only a *Southgate.* Mansfield (Not) and Peterborough (Cam) both have a *Westgate* but none of the other three. The latter's name means 'west street' although *Northgate, Eastgate* and *Westgate* in Lincoln, north of Peterborough, are all streets named for city gates.

Streets may be not only broad or narrow but long or short. Names denoting length include those of *Long Street,* Devizes (Wil), leading south out of the town from High Street, *Long Street,* Sherborne (Dor), dating from the 14th century, *Long Lane,* Bermondsey, south London, which at one time linked the village of Bermondsey with Southwark, and *Long Causeway,* Peterborough (Cam), although this last name is not recorded before the early 19th century. Hull (Hum) has a *Short Street* which dates from the same recent period. More originally, *Endless Street,* Salisbury (Wil), is so named because it formerly led out beyond the northern limits of the city. The name *Endless Way* similarly existed at one time in Cambridge and Ely (Cam).

Some streets are cul-de-sacs, with no through way. Their names may denote this, such as *Turnagain Lane* off Farringdon Street in the City of London, which ended at the River Fleet as a blind alley. A street of identical name and nature exists in Abingdon (Oxf). The name of *Blythe Road* by the Olympia exhibition centre in west London is said to be a corruption of *Blind Lane,* and an 18th-century map shows it ending simply in fields.

Almost the exact opposite of a cul-de-sac is a crossroads, where many roads meet to give access to one another. Oxford's famous *Carfax* is such a crossing, at the junction of the High Street, St Aldate's, Queen Street and Cornmarket. The name originated through French (modern *carrefour*) from Latin *quadrifurcus,* 'four-forked', referring to these four streets. Exeter (Dev) also once had its *Carfax,* at the junction of North Street, South Street, High Street and Fore Street, and a crossing of the name exists still today in Horsham (WSx) and Cannock (Sta). One better numerically than this is the crossroads (now a roundabout) known as *Five Ways*, Birmingham. The name dates from at least the 16th century.

Streets meeting at a central city crossroads are usually straight. Some streets are anything but, however, and have a name to indicate their irregular course. Two examples are *Corkscrew Hill*, West Wickham, and *Zigzag Road*, Kenley, both in south London. Another *Zigzag Road* is in Portland (Dor). *Crink Lane* in Southwell (Not) is a name that means 'crooked lane', while *East Bight*, Lincoln, means 'eastern angle'. The latter name originally referred to the land in a bend of the Roman wall here, however, rather than to the street itself. A more straightforward and common name is that of *Winding Road*, Halifax (WYo), earlier *Winding Lane*, which curves westwards into Northgate at its northern end. London's *Bow Street* belongs here, as its name refers to its curving shape. The 'bow' is now slight, however, as the street runs south from Long Acre into Wellington Street,

A common function of a street is to join two others. *Cross Street* may indicate such a link, as in Hull (Hum), where the street leads from West Street to Mill Street. However, this particular name could relate to an actual cross, so can be ambiguous. An alternative name for such a street is *Union Street*, which in Hull links Albion Street and Baker Street. The same city has *Great Union Street*, built in the early 19th century to join Witham Street and Clarence Street, and called *Great* to distinguish it from other Union Streets. Even here, though, there can be cause for ambiguity, since *Union Street* may relate not to a road junction but to a union workhouse, as it does in Hemel Hempstead (Her) (see Chapter 8, p. 115).

The function of a particular street may be contained in its name in different ways. *Warple Way*, Acton, west London and *Worple Road*, Isleworth, north London have identical names showing that they were originally bridleways. The name derives from conjectural Old English *werpels*, 'path', itself a derivative of *wearp*, meaning 'something thrown up' and referring to earth banked up at the sides when the track was levelled. Another name of this type is *Waingate*, Sheffield (SYo), meaning 'wagon street', but dating only from about 1700. *Packhorse Lane*, Loughborough (Lei), was the route by which timber was carried by packhorse from Charnwood Forest to Leicester. *Toll Gavel*, Beverley (Hum), was a road where tolls were collected. The unusual name is a combination of Old English *toll*, 'toll', 'tax' and *gafol*, 'tribute', 'rent'. More modern transport lies behind the name of *Tramway Avenue*, in Lower Edmonton, north London, and of *Tramway Path* along the western edge of Mitcham Common, south London.

Some streets run along (or even form) parish or other boundaries, and their names may indicate as much. *Boundary Street*, east London, marks the eastern boundary of the parish of St Leonard, Shoreditch, while *Boundary Road*, St John's Wood, northwest London, forms part of the northern boundary of the medieval manors of Lilestone and Tyburn. Today it is part of the boundary between the modern districts of Camden and the City of Westminster. *Westborough Road*, Maidenhead (Ber), formed the western boundary of the borough of Maidenhead until 1934, when the boundaries were extended. *Division Road*, Hull (Hum), marks the western boundary of the historic manor of Myton and the eastern boundary of the grasslands known as the Wold Ings, outside the medieval town. A road may equally follow a particular route without actually serving as a boundary in itself. One example is *Boutport Street*, Barnstaple (Dev), which was established just outside the medieval town walls. Its name derives from Old English *būtan porte*, 'outside the town'.

Any street named *new* would have been aptly if unoriginally descriptive at the time of its construction, although today many such streets are centuries old. There is a *New Street* in Birmingham, Coventry (WMd) and Salisbury (Wil), for example, and in these cities they date respectively back to the 15th century (when Birmingham expanded northwards), 14th century (when medieval Coventry began to grow to the east), and 13th century. Salisbury's *New Street* is thus one of the oldest in the city, and must have been named by contrast with an even earlier street, perhaps *Minster Street.* In the north of England York's *Newgate*, having the same meaning, dates from the 14th century, and the identically named street in Pontefract (WYo) dates also from this period. *New Street* in the City of London, however, is of a later date, and was so named in the 18th century when newly built as the reconstruction of a former alley off Bishopsgate. Oxford's *New Road*, west of the city centre, dates from the same century and is said to be the first new road to be built in the city since the time of William the Conqueror. *New York Road*, Leeds (WYo), has a rather misleading name, indicating merely that it was built as an extension of *York Road. Extension Street*, Ilkeston (Der), was constructed as an extension of *Market Street*.

A street with *hill* in its name is likely to be one that ascends (and descends), or to be elevated in some way. Names such as *Hill Brow*, *Hill Crest* and *Hill Rise* are frequently found, as are names ending in this word, such as London's *Herne Hill*, *Highgate Hill* and *Shooters Hill.* This last, in Greenwich, southeast London, should be distinguished from *Shoot Up Hill*, Camden, northwest London, which is so named because it is steep, and so 'shoots up'. Even here there can be exceptions, however. *Hill Street* in Poole (Dor) was originally *Hell Street*, a derogatory name rather than a topographical one. Better disguised is *Ber Street*, Norwich, which derives its name from Old English *beorg*, 'hill' (modern German *Berg*). The street runs along the top of a ridge to the south of the city centre. Much more obvious are Lincoln's *Steep Hill*, which with The Strait and the High Street represent Ermine Street as it crosses the city, and *The Mount*, Guildford (Sur), formerly *Mount Street*, a steep street leading southwest from the city centre up to the Hog's Back (see Chapter 2, p. 24). For examples of hill names as field names, see Chapter 4, p. 53.

A street named after a *bank* may or may not be elevated, depending whether it lies on or near the bank and on whether the bank is itself raised to any degree. *Bank Street*, Hull (Hum), is named after the nearby Derringham Bank, and London's

Bankside runs by the Thames on the south side of the river, next to what was originally a medieval embankment. Opposite it, on the north bank of the Thames, lies the similarly (but more royally) named *Victoria Embankment*, often referred to simply as *The Embankment*. To the west of it run *Millbank* and the *Chelsea Embankment*. All these London streets are elevated inasmuch as they are raised above the level of the river. Birmingham's *Digbeth* may belong here, if its name represents Old English *dīc-bæth*, as sometimes explained. The meaning is literally 'dyke-bath', referring to a pool by a causeway. The road runs southeast from the city centre over what was at one time a marshy area. A name similar to this is that of *Castle Dyke*, Stamford (Lin), a street to the east of the remains of the Norman castle on the site of its defensive embankment.

A *causeway* is by definition a raised road or path over marshy or damp land, properly one paved with limestone (Latin *calx*). Examples of such roads are *Fen Causeway* and *Maids Causeway*, Cambridge, both extending over low-lying ground by the River Cam, London's *Newington Causeway*, leading over marshy ground from Southwark to Newington as a continuation of Borough High Street, *Long Causeway*, Peterborough (Cam), running north from the city centre by a stream whose course has now long been diverted, and *The Causeway*, Petersfield (Ham), as what is now the A3 to Portsmouth to the south of the town. (For the origin of *Maids Causeway*, see Chapter 7, p. 104.)

Streets with names based on a form of *hollow* will conversely be in a dip or depression of some kind. *Holloway* is frequently found, as in north London, while *Holgate Road*, York, was originally a 'street in a hollow'. Nottingham has *Hollowstone Road* as a road that was originally a 'hollow way through rock', while Birmingham's *Holloway Head* is at the head (top) of the 'hollow way' running southwest out of the city centre to Edgbaston. The road is on sandy soil, and so would originally have been subject to continual erosion.

A low-lying road may have a form of *bottom* behind its name. One such is *The Bittoms*, Kingston upon Thames, southwest London, which was originally a piece of land in an angle between the Thames and Hogs Mill River.

If not noticeably elevated or depressed, a street may be simply level. *Wandsworth Plain*, southwest London, is a short level street off Wandsworth High Street at the top of nearby Richmond Hill, while in Oxford *The Plain* is an open, level road junction at the southern end of Magdalen Bridge. It is located on the site of the medieval St Clement's Church, demolished in 1830.

The nature of the ground or soil on which a street or road is established can also feature in its name, especially if the surface is unusually stony or soft. *Chiswell Street*, Finsbury, north London, runs over stony soil, and derives its name from Old English *ceosol*, 'flint', 'gravel' (as for Dorset's well-known *Chesil* Beach). More commonly, a street name will be based on a form of *stone* to indicate its hard surface. *Stoney Street*, Nottingham, is one such, recorded as *Stanstrete* in the early 14th century. *Stonegate*, York, is another, and dates from the 12th century (as *Stangate*), *gate* here having the common northern sense of 'street'. Both these streets would have been paved with stones. *The Steyne*, meaning 'stony place' (not necessarily paved), is the name of a street in west London, while Brighton (ESx) is famous for its *Old Steine*, a name of identical origin. (Streets called *The Steyne* in Worthing (WSx) and Bognor Regis (WSx) are named after the Brighton original.) St Ives (Cor) has *The Stennack* as its equivalent, from the Cornish *stean*, 'stone'. *The Pavement*, York, however, is more obviously a stone-paved route, similar to the three streets of the name (*High*, *Low* and *Middle*) in Nottingham.

Greenway Street, Chester, and *Grassy Lane*, Maidenhead (Ber), have names referring to their grassy or grass-lined routes. *Clay Street*, Colchester (Ess), is self-descriptive, as are *Marsh Avenue*, Mitcham, southwest London, and *Marsh Street*, Bristol. The equivalent name in Manchester and elsewhere in the northwest of England is *Moss Lane* (see the Appendix, p. 197).

SELF-DESCRIPTIVE NAMES

Many colourful names have evolved to refer to the muddy surface of a road or street, such as *Featherbed Lane*, Hemel Hempstead (Her), or *Honeypot Lane*, Stanmore, northwest London. *Honey Street*, Hertford, was at one time equally muddy or sticky, as probably was London's *Pudding Lane. Summer Lane*, Birmingham, also belongs to this category, since the name implies that the route was impassable in winter. *Porridge Pot Alley*, a footpath leading up from the River Wey in Guildford (Sur), may similarly denote the former miry nature of the terrain here.

Some such names conceal their unpleasant meaning in modern terms. *Full Street*, Derby, has a name that means 'foul street', while *Lurk Lane*, Beverley (Hum), should be understood as 'dirty lane', as it derives from Old English *lort*, 'dirt', 'mud'. *Addle Street*, in the City of London, was also muddy and dirty, taking its name from Old English *ādela*, 'mud', 'filth' (the source of modern *addle*). It may have acquired its unpleasant state from the continual passage of cattle, with their residue of stinking dung and urine. Oxford's *Squitchey Lane*, on the other hand, although popularly regarded as meaning 'squelchy' or 'squishy', probably took its name from the *squitch* or quitch grass that grew there.

Other names of this type are more generally disapprobatory. *Farthing Lane*, Watford (Her), is probably a contemptuous nickname for an insignificant route, as is *Pennyfarthing Street*, Salisbury (Wil). The latter was also the former name of *Pembroke Street*, Oxford, and *Botolph Lane*, Cambridge, respectively. On the other hand *Halfpenny Lane*, Pontefract (WYo), originally *Penny Lane*, may simply have referred to the cost of the toll payable.

Friday Street, Henley-on-Thames (Oxf), is an example of a street name found elsewhere. It implies poverty, as Friday was regarded with ill favour in medieval times. Not only was it the day of the Crucifixion, but (in observance of this religious event) was a fastday.

Hockley, Nottingham, may well have been a place of ill repute. It probably adopted its name from *Hockley-in-the- Hole*, London, notorious in the 17th and 18th centuries. This in turn seems to have taken its name from *Hockliffe* (Bed), which was known as *Hockley in the Hole* for its location in a depression. It was noted for its robberies. A similar name is that of *Hustlergate*, Bradford (WYo), which derives from the hustlers or pickpockets who frequented it likewise in the 18th century. In the same category is *Shootersway*, Berkhamsted (Her), which means 'robbers' way', from Old English *scēacere*, 'robber'. This same word also gave the name of the *Sugar Way* instanced in Chapter 2.

Rotten Row is undoubtedly a derogatory name, although the precise sense remains disputed. It is found in London's Hyde Park and in Lewes (ESx), among other places, and was the former name of Nottingham's *Cheapside*. It may mean 'rat row', as a term of contempt, or else refer simply to the 'rotten' or poor soil. It is not likely to be a corruption of French *route du roi*, 'king's way', as popularly explained for the famous London track for horseriders. (Its surface is sandy, so that the soil is indeed 'rotten' rather than firm.)

Cold Harbour (or *Coldharbour*) implies an inhospitable place. The name means literally 'cold shelter' and is found, for example, in Sherborne (Dor) and Poplar, east London, as well as widely elsewhere, often as the street name *Coldharbour Lane*.

Not all such judgmental names are negative, however. *Prospect Place*, Hull (and elsewhere), implies a street with an agreeable view or vista, as does the same city's *Belle Vue Terrace*. It is possible that *Beulah Hill*, Croydon, south London, has a name that means 'beautiful spot' if it represents Old French *beau lieu*. Perhaps *Peep-o-Day Lane*, Abingdon (Oxf), may have been intended as an approbatory name, describing an east-running street with a view of the rising sun. On the other hand, *Mount Pleasant*, though a fairly common street name, can be used of a place that is far from pleasant. For a consideration of its meaning, see Chapter 12, p. 176.

Colourful names are likewise often found for a distant or remote street. There may be a reference to a far-off geographical place, as for *Gibraltar Walk* or *Virginia Road*, both in Bethnal Green, east London. *Worlds End* is a stock name for a remote place, as for the street so called in Chelsea, west London, which dates back to the 17th century, when it lay at the western extremity of what was then the village of Chelsea. However, *Fargate*, Sheffield (SYo), is less romantic, simply meaning 'far street'. It is now virtually in the city centre, but in the late 17th or early 18th century, when it is first recorded, it applied to a southward extension of the High Street towards the moor.

Street names referring to trades and occupations are considered in Chapter 7, p. 88. It is therefore appropriate to include here names that relate to play rather than work. 'Play' chiefly implies sports and games, which in some cases were actually played on the site of the street itself. Examples are *Bowlalley Road*, Hull (Hum), and *Bowling Alley Walk*, Dorchester (Dor), *Buttgarden Street*, Bideford (Dev), relating to archery, also *The Butts* in several towns, such as Brentford, west London, and Warwick and *Trippet Lane*, Sheffield (SYo), from a North Country game similar to tip-cat.

Two of the best known 'street sport' names are London's *The Mall* and *Pall Mall*, both referring to pall-mall, a game similar to croquet in which a ball was driven by a mallet along an alley and through an iron ring. (The ultimate origin of the word is in Italian *pallamaglio*, from *palla*, 'ball' and *maglio*, 'mallet'.) *Pall Mall* dates from the mid-17th century. After the Restoration, Charles II formed a new pall-mall alley in St James's Park parallel to the original one, and this is *The Mall*, now a famous processional avenue. A less energetic although equally fashionable recreation lies behind the name of *Gentleman's Walk*, Norwich. It was originally *Le Nether Rowe*, but gained its present name in about 1720. A contemporary account explains that 'on Saturdays, at Sessions time & other occasions, when the Gentlemen of the Country resort to ye City, this is the usual place of their walking for their divertion & having discourse together, as it is also for the Principal Citizens' (J. Kirkpatrick, *The Streets and Lanes of the City of Norwich*, *c.*1720). Akin to this is *Fetter Lane* in the City of London, which takes its name from the *fewters* or idlers who strolled or lounged there. (The word is more correctly *faitour*, an obsolete term for an impostor or sham vagrant, from Old French *faitor*, 'doer'.)

Love Lane is a common street name relating to a more universal diversion. Towns and cities with such a street include Chester, Loughborough (Lei), Pinner, northwest London, Salisbury (Wil), Sheffield (SYo), Stepney, east London and Weymouth (Dor). The name usually dates from the 16th or 17th century, and apart from the obvious activity also implied a street that was dark or dangerous. *Maiden Lane* or *Maiden Street* usually refers more explicitly to prostitutes, and is likewise found in several towns, including London, Nottingham and Stamford (Lin). In some cases, however, the name may be genuinely innocent, simply referring to a street where girls and young women gathered, or one near a nunnery. This appears to be the case for *Maiden Street*, Weymouth (Dor), which dates from the 14th century. In other instances the name may refer to an inn called the Maid's Head or something similar. Two streets that seem to have a similar origin are *Mabgate*, Leeds (WYo), and *Meddon Street*, Bideford (Dev). The former, first recorded in the late 15th century, appears to derive its name from *Mab*, a pet form of the female personal name *Mabel*, here used generically for a 'loose woman'. The latter name, not necessarily referring to prostitution, evolved from *Maydenstrete*, recorded in the 14th century.

Finkle Street is now believed to have approximately the same connotations as *Love Lane*. It was long thought to have derived either from a Scandinavian personal name, from a northern dialect word *fenkl*, meaning 'bend' (related to modern German *Winkel*, 'angle', 'corner'), or from a dialect word *fenkel* meaning 'fennel'. It is found in many northern towns, including Carlisle (Cum), Kendal (Cum), Knaresborough (NYo), Lincoln (formerly), Ripon (NYo), Pontefract (WYo), Workington

(Cum) and York. Norwich, too, has its *Finkelgate.* Most toponymists are now inclined to accept the origin proposed by Eilert Ekwall in *Etymological Notes on English Place-Names* (1959), where he derives the name from a conjectural Middle English word *finkle* meaning 'to pet', 'to cuddle'.

The oldest and most explicit of all street names of this type has survived in *Grope Lane*, Bristol, *Grape Lane*, York and *Grape Passage*, Reading (Ber). The name is a euphemistic short form of *Gropecuntelane*, and this (or a form of it) was formerly found in several towns. York's *Grape Lane* was thus *Grapecuntlane* in the 14th century, while a hundred years earlier the City of London had its *Gropecontelane* and Oxford its *Gropecuntelane.* The 'four-letter word' that lies at the heart of the name is given its earliest citation by the *Oxford English Dictionary* (which admitted it to its pages only in 1972) from its occurrence in the Oxford street name, which dates from about 1230. The street in that city that bore it is now *Magpie Lane. Grope* here relates to feeling one's way in the darkness of the lane as much as to any physical amorous advances.

Silent Street, Ipswich (Suf), is a name which, if it means what it says, could perhaps relate to the straw laid in the street to deaden the sound of cartwheels and horses' hooves by a nearby hospital. The name has not been recorded earlier than 1764, but seems to have been well established at this time. In Saffron Walden (Ess), there is *Sound Alley*, which is said to derive from the echoes which can be heard down this narrow lane. *Manet Street*, Hull (Hum), is said to derive its name from Latin *manet*, 'it remains', referring to a permanent way. Such a name is markedly pedantic when compared to *Tink-a-Tank*, Warwick, a narrow path by St Mary's Church. It was originally *Cannon Row*, but its present name is more imaginatively said to represent the echo of passing footsteps.

4. FIELD AND WATER NAMES

Names of existing or former fields are frequently adopted for a road or street laid out across or near such a field. The name may actually contain the word *field* as an indication of such a borrowing, although there are undoubtedly many streets containing this word purely as a pleasant-sounding rural term, with no genuine reference to an actual field.

Examples of adoptions from genuine fields are the names of *Barnfield Road*, Edgware, north London, *Beanfield Road*, Sawbridgeworth (Her), *Beechfield Road*, Doncaster (SYo), *Broadfields Avenue*, Edgware, *Broomfield Road*, Ealing, west London, *Buryfields*, Guildford (Sur) (ie 'borough fields'), *Churchfield Road*, Acton, west London, *Courtfield Gardens*, Earls Court, west London, *Highfield Avenue*, Roe Green, northwest London, *Linkfield Road*, Isleworth, north London (the reference is to *linches*, boundary ridges), *Mayfield Road*, Hackney, east London, *Oakfield Road*, Palmers Green, north London, *Pitfield Street*, Shoreditch, north London (named for its excavation), *Ryefield Avenue*, Hillingdon, west London, *Smithfield Street*, City of London (where the 'smooth field' became a famous meat market), *Southfield Road*, Acton, west London, *Springfield Garden*, Ilkeston (Der) (so named for the many springs here) and *Woodfield Road*, Ealing.

Names containing *mead* or *meadow* are also often borrowed from those of fields. Examples are *Broadmead*, Bristol, *Castle Meadow*, Norwich, *Meadow Lane*, Loughborough (Lei), *Moggs Mead*, Petersfield (Ham) (apparently after the field's original owner) and *Westmead Road*, Sutton, south London.

Names containing *croft* similarly relate to a field, properly a small enclosed one. Examples include *Bancroft Lane*, Mansfield (Not) (ie 'bean field'), *Croft Street*, Lincoln, *Greencroft Street*, Salisbury (Wil) (there is still a grassy area here called *Green Croft*), *Haycroft Road*, Stevenage (Her) (the name dating from the 14th century), *Northcroft Lane*, Newbury (Ber) (the town meadow, by the River Kennet), *Peachcroft Road*, Abingdon (Oxf), *Tentercroft Street*, Lincoln (where newly woven cloth was stretched on *tenters*) and *Turncroft Lane*, Stockport (GtM) (ie 'round field'). *Peascod Street*, Windsor (Ber) was originally *Puscroftestrete*, 'street by the field of peas', with *peos-croft* popularly understood (or misunderstood) as *peas-cod*, 'peapod'.

Terms also denoting different types of field occur in such names as *Applegarth Lane*, Bridlington (Hum) ('apple orchard', with *garth* from Old Norse *garthr*, 'enclosure'), *Forlease Road*, Maidenhead (Ber) (ie 'fields in front', from Old English *fore* and *lǣs*, 'pasture'), *Gerston Road*, Paignton (Dev) (from Old English *gærs-tūn*, 'grass enclosure'), *The Green*, Sherborne (Dor), *Hall Ings*, Bradford (WYo) (from Old English *eng*, 'meadow'), *The Ham*, Brentford, west London (from Old English *hamm*, 'enclosure', 'water-meadow'), *Long Leys Road*, Lincoln (from Old English *lēah*, 'meadow', modern poetic *lea*), *Northernhay Street*, Exeter (Dev) (from Old English *gehæg*, 'fence', 'enclosure', related to modern *hedge*), *Padholme Road*, Peterborough (Cam) (ie 'toad meadow', from Old Norse *padda*, 'toad' and *holmr*, 'water-meadow'), and *Voundervour Lane*, Penzance (Cor) (ie 'lane by the big meadow', from Cornish *bounder*, 'pasture' and *meur*, 'big'). York's *Peaseholme Green* ('green by the water-meadow where peas are grown') has the same meadow term as *Padholme Road*, and Bristol's *Pithay* ('pit enclosure') has the same word as that in *Northernhay*.

Some streets are named from field names that relate to size, such as *The Half Acre*, Brentford, west London, *Halfhide Lane*, Cheshunt (Her) and *Long Acre*, running from Leicester Square to Drury Lane in central London. A *hide* was the term for a plot of land sufficient to support a single family, and lies behind the name of London's famous *Hyde Park*. Other field names of less direct or even obscure origin include *Curtain Road*, Shoreditch, north London (where coincidentally a Curtain Theatre was built, although the name predates this), *Fobney Street*, Reading (Ber) ('Fobba's island') and *Gresty Road*, Crewe (Che) ('badger run', from conjectural Old English *grǣg*, 'badger' and *stīg*, 'path'). London's *Soho Square* is named for the surrounding area, itself apparently so called from the hunting cry that was heard here before the fields were built over in the late 17th century. ('So-ho' was the cry for a sighted hare, while 'Tally-ho' was its equivalent for a fox.)

Names can also be adopted from gardens, often implying vegetable gardens or orchards, such as *Cross Deep Gardens*, Twickenham, southwest London (referring to a pool by the Thames here), *Garden Avenue*, Ilkeston (Der), *Nursery Street*, Sheffield (SYo) (so named from a garden here known in the mid-18th century as *The Nursery*), *Orts Street*, Reading (Ber) (from Medieval Latin *ortus*, 'garden', related to modern English *horticulture* and *orchard*), *Saffron Hill*, Holborn, northeast London (from a garden where saffron was grown in the grounds of Ely House) and *Strawberry Street*, Hull (Hum) (built over strawberry fields and tea gardens). London's *Spring Gardens*, however, was originally a pleasure ground laid out in a park. *Cherry Street*, Birmingham, is named from a cherry orchard, as is the more originally named *Cherry Bounce*, Hemel Hempstead (Her). The second word of the latter name is said to refer to the motion experienced by coach passengers riding along this poorly made road. *Plantation Road*, Oxford, has a self-descriptive name.

Paradise Street is a fairly widespread name. In some cases it may simply be a generally agreeable name, in which case it belongs in Chapter 12, not here. But in certain instances it specifically relates to a garden or orchard, a 'garden of Eden'. This is so in Oxford, where *Paradise Square* dates from the 17th century and refers to the orchard that belonged to the Penitentiary Friars. It may not be irrelevant that Church Latin *paradisus* could actually mean 'enclosure'. It is likely that *Paris Street*, Exeter (Dev), has this particular origin, and similarly relates to an enclosure. It gained its present form through Old French *pareis* and under the influence of the name of the French capital.

Some street names refer to a nearby *park*, as for *Park Street*, Birmingham, running along the west side of the *Little Park*, and *Parks Road*, Oxford (*Park Street* in the 1830s), from the University Parks.

A name with *hill* may also be a transferred field name, or else one adopted from that of the hill itself. Examples include *Dockin Hill*, Doncaster (SYo) ('dock hill'), *Green Hill*, Sherborne (Dor), *Moffy Hill*, Maidenhead (Ber) (apparently 'mossy hill', with the old 'long' *s* read as *f*), *Muswell Hill*, north London ('mossy spring hill'), *Romeland Hill*, St Albans (Her) ('empty land hill', the first part of the name representing Old English *rūm*, 'spacious', 'empty'), *Seed Hill Road*, Huddersfield (WYo) ('hill seeded with corn'), *Shude Hill*, Sheffield (SYo) ('pod hill', from Old English *scēod*, 'pod', probably denoting a field where corn was threshed), *Snig Hill*, Sheffield (referring to a hill where a cart had to be braked with a pole or *snig*) and *Windhill*, Bishop's Stortford (Her) ('windy hill'). The word for 'hill' (Old English *hyll*) is disguised in *Royal Lane*, Hillingdon, west London (ie 'rye hill') and in *Rynal Street*, Evesham (H&W) (probably 'ruin hill').

Other words for different types of hill can be seen behind such names as *Blackborough Road*, Reigate (Sur) ('road by the barren hill', with *borough* representing Old English *beorg*, 'hill', and *black* denoting barrenness), *Downage*, Hendon, northwest London ('hill with a hedge', from Old English *dūn*, 'hill', 'down' and *hecg*, 'hedge') and *Low Fellside*, Kendal (Cum) (with *fell* from Old Norse *fjall*, 'hill', 'mountain'). *Beaumont Street*, Oxford, although meaning literally 'beautiful hill', takes its name from the former Beaumont Palace here, built by Henry I in the 12th century outside the town gate.

Woods also give the name of streets, whether the actual word *wood* appears or not. One thus finds *Bradshaw Way*, Derby ('broad copse', from Old English *brād*, 'broad' and *sceaga*, 'copse'), *The Grove*, Dorchester (Dor), *Nutholt Lane*, Ely (Cam) ('nut wood lane') and *Sheethanger Lane*, Hemel Hempstead (Her) ('lane by the projecting piece of land at the wooded slope', from Old English *scēat*, 'corner of land' and *hangra*, 'hillside wood', literally one that 'hangs').

Old English *lēah* meant both 'wood' and 'clearing in a wood', and is commonly found as the *-ley* in many place-names, so that a street called *Bradley Lane*, for example, could mean 'lane by a broad wood' or 'lane by a broad clearing'. It is found in the name of *Whalley Hayes*, Macclesfield (Che), where the first word thus means 'wheel clearing', while the second word means 'hedges'. The street called *The Ridings* in Ealing, west London, is similar, since it derives from another word for a clearing, conjectural Old English *ryding*, literally 'land where the trees have been got rid of'.

Land itself occurs in many field names transferred to streets. Among examples are *Bearland*, Gloucester ('bare plot of land'), *Chanterlands Avenue*, Hull (Hum) ('land providing income for chantry priests'), *Lands Lane*, Leeds (WYo) ('lane by the land of the common field'), *Newland*, Lincoln and *Newlands*, Northampton, as well as widely elsewhere ('new land'), *Norland Road*, Chelsea, west London ('northern land', here the northern portion of the manor of Abbot's Kensington), *Scotland Street*, Sheffield (SYo) ('street by the taxed land', from Middle English *scot*, 'tax', 'payment') and *Shireland Road*, Birmingham ('road by the land between the shires of Warwick and Stafford', ie land by the stream that formed the boundary here). *Lagland Street*, Poole (Dor), was originally *Lage Lane*, so means 'lane by the marsh', not 'marshland', from conjectural Old English *lagge*, 'marsh'. *Lots Road*, Chelsea, west London, ran over land that was allotted annually at Lammastide.

Aynam Road and *Little Aynam*, both in Kendal (Cum), have names referring to land that had been cut off from an estate, from Old Norse *af-nám*, 'detached land' (related to modern German *abnehmen*, 'to remove', literally 'take off'). *Glebe Road*, Dagenham, east London, relates to the *glebe*, or land allotted to the local parish church for the use of its clergy. *Kensington Gore*, west London, is named for land in a triangular plot, from Old English *gāra*, 'gore'. The particular wedge of land here was that bounded by Queen's Gate, Cromwell Road and Kensington Road. As it is now current, the name applies to the street that extends westwards from Kensington Road. *Longdales Road*, Lincoln, has a name that means 'road by the long doles', that is, the long shares of land. *Longshut Lane*, Stockport (GtM), means 'lane by the long corner of land', with *shut* representing Old English *scēat*, 'corner of land' (as for *Sheethanger Lane*, above). More generally, the name of *Morrab Road*, Penzance (Cor), relates to the seaward part of the parish here, deriving from conjectural Cornish *morrep*, 'seashore', as distinct from the *goonran*, the upland part of the parish. *Shoe Lane*, Westminster, west London, takes its name from a field or portion of land that was originally shoe-shaped.

Names derived from plants and trees may either be field names or else relate to prominent or profuse (or, conversely, rare) instances of the particular plant or tree. *Crabtree Lane*, Hammersmith, west London, was originally a field name, recorded as *Crabbetrehalfacre* in the mid-15th century. Similarly *Elder Way*, Chesterfield (Der), was *Ellar Yard* in 1692, meaning 'elder-tree enclosure'. But *Acacia Avenue*, wherever it occurs, may originally have simply denoted this particular tree, without deriving from a field name. The name has become symbolic of suburban gentility. Similar names include *Elm Street*, found in London and elsewhere, *Fennel Street*, Loughborough (Lei), *Flag Lane*, Crewe (Che) (for a lane abundant in reeds), *Holly Drive*, Maidenhead (Ber), *Mulberry Walk*, Chelsea, west London (although a silkworm nursery was nearby), *Oak Street*, Norwich (from a large hollow oak that stood in a nearby churchyard) and *Wicken*, Sheffield (SYo) (for its willows). It seems likely that *Spen Lane*, York, was named for its aspens. Sheffield's *Balm Green* is perhaps named for its balsam or balm-mint.

Names relating to wells or springs may equally have been transferred to a street from their use as a field name. There is a healthy representation of such names. They include *Annetwell Street*, Carlisle (Cum) (perhaps originally meaning 'St Ann's well'), *Goswell Road*, Finsbury, north London ('goose spring'), *Keldgate Road*, Beverley (Hum) ('street with a spring', from Old Norse *kelda*, 'spring' and *gata*, 'street'), *Newell*, Sherborne (Dor) ('new spring'), *Pearl Street*, Macclesfield (Che) ('street with a bubbling spring', from a word related to modern English *purl*),

Redwell Street, Norwich (from a well that perhaps had a red frame or cover), *Springs*, Wakefield (WYo) (a name dating from the mid-15th century), *Wellgate*, Rotherham (SYo) ('street by a spring'), *Well Lane*, Lincoln and *Well Place*, Cheltenham (Glo).

Similar names derive from local ponds or lakes. Among them are *Drypool Way*, Hull (Hum) (by a pool that had dried up), *Flash Lane*, Enfield, north London (from Middle English *flasshe*, 'shallow pool'), *Latchmere Road*, Battersea, south London (from Middle English *lache*, 'stream' and Old English *mere*, 'pool'), *Meer Street*, Stratford-on-Avon (War) (also from *mere*, 'pool'), *Muspole Street*, Norwich ('street by the muddy pool', from Old English *must*, literally 'must', 'grape-juice', implying cloudy water, and *pōl*, 'pool') and *Pond Place*, South Kensington, west London (the pond is now long filled in). As ever, names of this kind can be heavily disguised. Two examples are *Drummer Street*, Cambridge and *Peerless Street*, Finsbury, north London. The former takes its name from a pond recorded in the mid-13th century as *Drusemere*, probably meaning 'muddy pool', from Old English *drosn*, 'dregs', 'dirt' and *mere*, 'pool'. The latter is apparently a euphemistic reference to a former *Perilous Pond*, said to be so called because boys or young men had been drowned when swimming in it.

Streets named after individual rivers are considered in Chapter 5, p. 67. There are several streets that are named more generally after a local watercourse or channel, however, from a small stream to a large river or waterway. They include *Conduit Street*, Abingdon (Oxf) and elsewhere, for a nearby piped water channel, *Lavant Street*, Petersfield (Ham), from a local word for an intermittent stream (not the small River Drum that flows through the town), *Leads Road*, Hull (Hum), for a local canal or drain (Old English *gelād*, 'watercourse', modern *lode*), *New Canal*, Salisbury (Wil), from a cut of the River Avon that existed here until the 19th century, *Pipe Lane*, Bristol, for a conduit, and the ubiquitous *River Lane* and *Water Street*, alongside a river or other watercourse. *Waterside* itself is also fairly common, as in Stratford-on-Avon (War) (by the Avon), and Knaresborough

(NYo) (by the Nidd). London's famous *Strand* originally crossed the northern *strand* (shore) of the Thames, and the river was much wider before it was contained by embankments in the 19th century. Today the Strand is almost a quarter of a mile from the river, even at its closest point (at the upper end of Lancaster Place, leading to Waterloo Bridge). *Scattergate*, Appleby (Cum), also belongs here. The name refers to a street by an open sewer, from a word related to modern *shit. Gate* here means 'street', as often in northern names.

Any names with *ford* also imply a watercourse, or the crossing of one, as for *Ford Street*, Derby. (The name relates not to the Derwent, the city's main river, but to a crossing of the much smaller Markeaton Brook.) *Swadford Street*, Skipton (NYo), probably means 'track to the ford', from Old English *swœth*, 'track' (modern English *swathe*). Other names relating to water include *Floodgate Street*, Birmingham, by the River Rea, *Reservoir Road*, Ruislip, west London, leading to Ruislip Lido, and *Spa Road*, Bermondsey, south London, referring to the chalybeate spring discovered in tea gardens here in the late 18th century. This last name is found elsewhere in spa towns, or towns that once had spas or mineral springs, such as Melksham (Wil).

As for watercourse names, street names containing names of animals and birds may equally have originated as field names. *Bollo Lane*, Acton, west London, has a name meaning 'bull hollow', from conjectural Old English *bula*, 'bull' and *holh*, 'hollow'. *Bull Green*, Halifax (WYo), has a similar name. This same city has *Cow Green* as a complementary name. More indirectly, *The Calls*, in Leeds (WYo), was originally a lane where cattle were 'called', ie driven. *Catte Street*, Oxford, was frequented by cats, presumably of the 'feral' kind. The name dates back to the 13th century. In the 15th century a hall named St Catherine's stood in the street, and in the 19th century the street name was altered to *Catherine Street* on the assumption that *Cat* was a short form of this personal name. After local agitation, including a plea from the Poet Laureate Robert

Bridges, who lived in Oxford, the old name was restored in 1930, but in the medieval spelling.

Colley End Road, Paignton (Dev), relates to an enclosure where pigeons were kept. The name evolved from *Culverheye Strete*, from Old English *culfre*, 'dove', and *gehæg*, 'enclosure'. *The Coneries*, Loughborough (Lei), is a name referring to a rabbit warren, from Middle English *coninger* (*coni*, modern *cony*, being 'rabbit' itself). *Dear Hay Lane*, Poole (Dor), refers to a nearby deer enclosure, while *Daw Bank*, Stockport (GtM) is named for jackdaws. Bristol's *Frog Lane* has a self-explanatory name, but *Golden Square*, Westminster, west London, is of less obvious origin. It probably refers to geldings once kept here. *Herne Hill*, Lambeth, south London, refers to the herons that must have been frequently seen on the marshland here. *Horse Castles*, Sherborne (Dor), appears to refer to an ancient earthwork where horses were kept. *Hounds Gate*, Nottingham, tells of a street frequented by dogs, or one where they were kept. A name of identical origin is that of *Hungate*, found in Lincoln and elsewhere, for example in Market Weighton (Hum). *Kitt Hill*, Sherborne, relates to kites (the bird), not kittens, while *Lambhay Street*, Plymouth (Dev), refers to a local enclosure for lambs. *Ramshill*, Petersfield (Ham) may mean what it says, although a derivation from Old English *hramsa*, 'wild garlic', is also possible. Nottingham's *Hounds Gate* is crossed by *Spaniel Row*, a name dating from the 15th century and said to refer to the king's spaniels, which were housed here. A similar name is that of *Dog Kennel Lane*, Southwark, south London, where kennels owned by Prince George of Denmark are traditionally said to have been sited. *Swan Street*, Loughborough, is named from a former nearby swannery (elsewhere it might have been named from an inn), while in Plymouth *Vauxhall Street* is a name corrupted from *foxhole*, doubtless under the influence of the famous London pleasure gardens. More generally, *Stott Hill*, Bradford (WYo), takes its name from Old English *stot*, a word meaning a draught animal of some kind, such as a horse or ox. Finally, London's former *Grub Street*, now Milton Street, was probably so named for its infestation of maggots, although some see an origin in a personal name Grub or Grubbe. Today the name is associated with the literary hacks and impoverished writers who eked out a living there.

Street names involving animals more explicitly for trading purposes, typically in a market, are considered in Chapter 7.

5. DIRECTIONAL NAMES

A directional name is one deriving from the destination of a road or street, that is, from the place or object to which it leads. The name may be precise and specific, such as the *London Road*, or more general, as the *Great North Road*. Several of the names quoted in the preceding chapters have been of this type.

A directional name may not be immediately apparent. Many streets with names seeming to derive from their destinations have actually been adopted from a personal name. This is noticeably the case in a large city, such as London or Birmingham, where (say) *Newton Road* may lead, or have at one time led, to a place called Newton but which could equally have been named for someone of this surname. As a result the converse can apply, so that a name actually deriving from that of a place may be supposed to derive from that of a person, or at best there may be uncertainty with regard to the origin, especially where the destination has an unfamiliar name, or one altered from the original. Where a directional name does derive from a place-name, it can equally appear to refer to some well-known town of the name instead of to an unfamiliar local village. Examples are *Swindon Road*, Cheltenham (Glo) and *Windsor Street*, Stratford-on-Avon (War) (see below).

It is therefore of value to consider some examples of directional names, and see what forms they can take.

Broadly speaking, such names fall into three groups: those named for a town or village, those named for some natural topographical feature such as a hill or wood, and those named for a building or other manmade object. The second group contains many streets named for a particular river; the third group, not surprisingly, includes several streets named for bridges over a river.

On the whole, a street with a name ending in *-gate* in the north and east of England (where it means 'way', 'street') will turn out to have a directional name. This is partly because of the basically 'static' implication of *street*, partly because Old Norse *gata*, which lies behind such names, has a basic sense of 'going' or motion that may still be intuitively sensed. (The word is directly related to English *gait*, perhaps also indirectly to *gate* in the sense of 'opening giving access to a place', and even possibly ultimately to *go*.)

Some examples of directional names of potentially misleading or ambiguous origin include: *Abington Street*, Northampton (to a former village of the name that is now a northeast district of the city), *Arden Street*, Stratford-on-Avon (War) (to Henley-in-Arden; the name complements that of Stratford's *Henley Street*), *Barnsley Street*, Hull (Hum) (leading to the railway to Barnsley), *Barton Street*, Gloucester (to Abbots and Kings Barton), *Chestergate*, Macclesfield (Che), also Stockport (GtM) (to Chester), *Dunham Road*, Altrincham (GtM) (to Dunham Massey; compare *The Downs*, below), *Erleigh Road*, Reading (Ber) (to the village of *Earley*, east of the city), *Fisherton Street*, Salisbury (Wil) (not to Fisherton de la Mere, the village some 10 miles northwest of the city, but to the former manor and village of Fisherton Anger, now absorbed into western Salisbury), *Franchise Street*, Weymouth (Dor) (to Little Francis, south of the town), *Great Dover Street*, Southwark, south London (to the Old Kent Road, and so ultimately to Dover (Ken)), *Isle of Wight Road*, Swanage (Dor) (obviously not to the island itself but to a view of the island), *King Street*, Norwich (to *Conesford*, one of the settlements that combined to form the original city), *Mytongate*, Hull (Hum) (to Myton, one of the settlements which combined to form medieval Hull), *Rutland Terrace*, Stamford (Lin) (to the historic county of this name), *Scotch Street*, Carlisle (Cum) (to Scotland), *Stricklandgate*, Kendal (Cum) (not to Great or Little Strickland, some 20 miles north of Kendal, or for the common Cumbria surname, but to Strickland Ketel, immediately north of the town), *Swindon Road*, Cheltenham (Glo) (not to Swindon (Wil) but to the identically named village north of Cheltenham), *Westwick Street*, Norwich (to Westwick, one of the original settlements that combined to form the present city; compare *King Street*, above), *Windsor*

Street, Stratford-on-Avon (not to the Berkshire town, or for the royal family, but to a piece of land outside Stratford).

In some cases a road that leads from London to a particular town some distance away may have an identical directional name repeated in the towns through which it passes. Thus the long-established road from London to Bath, now mainly the A4, is named as *Bath Road* in Hounslow, west London (as the A3006 leading to the A4), in Hillingdon as it runs past Heathrow Airport, in Slough (Ber), in Maidenhead (Ber), in Reading (Ber), in Newbury (Ber), in Hungerford (Ber) and in Chippenham (Wil). In all these towns it is thus an established individual street name. It is an old one, too, recorded in Chippenham as *Batheweye* in about 1300. Beyond Bath, the A4 continues to Bristol, and there, too, it is also naturally *Bath Road*.

Two interesting examples of directional names, both in London and both widely known, are *Oxford Street*, in the West End, and *Theobalds Road*, in Holborn, northeast of the city centre.

Oxford Street is indeed the start of the road to Oxford (the modern A40). It was actually known as *The Road to Oxford* in the late 17th century, and was referred to by Horace Walpole in a letter dated as late as 26 April 1771 as *Oxford Road*. But while this is undoubtedly the prime sense of the name, the present Oxford *Street* was established in the 1720s when Edward Harley, second Earl of *Oxford*, developed his estate here. The street thus takes its name both from its destination and from the aristocratic title of its prominent landowner. (See Chapter 13, p. 179, for further street names attributable to him and his family.)

Theobalds Road (originally *Theobalds Row*) has a name that appears to derive from one Theobald. It actually gained its name from the fact that it was the route along which James I rode on his journey from Whitehall to Theobalds (Her), where he had a hunting lodge. The original name is preserved in that of *Theobald Street* (Her), as the 2½-mile-long road that today follows the railway between Radlett and Borehamwood stations. *Theobald* literally means 'Titta's burst', that is, relates to a landslip or fissure in the ground owned by one Titta. The same Anglo-Saxon name is probably behind that of *Tykes Water*, the stream whose course Theobald Street also mainly follows. (*Tykes* appears to be a miscopying of *Tytes*.) The name later became falsely associated with the personal name *Theobald*.

Many of the field names, spring names and the like mentioned in Chapter 4 are also directional names. A street or road is after all desirable to give access to a field, spring or river. *Charnwood Road*, Loughborough (Lei), ran to Charnwood Forest. *Childwell Alley*, Colchester (Ess), led to the spring of this name ('children's spring'). Colchester has two other streets named for the springs to which they originally ran: *Stanwell Street* (to the 'stony spring') and *East* (and *West*) *Stockwell Street* (to a 'spring by a tree-trunk'). *The Downs*, Altrincham (GtM), led to Bowdon Downs, which gave the name of Dunham Massey here (see *Dunham Road*, above). *Fielden Way*, Warwick, led east to Ryton Fields, as against a western route to the Forest of Arden. *Glyde Path Road*, Dorchester (Dor), is more unusual in being a road that led to a path, in this case *Glyde Path*. (The name probably derives from Old English *glida*, 'kite', as these birds were frequently seen here.) Both the *Great North Road* and *Great West Road* (see Chapter 2, p. 28) run generally and respectively north and west from London. *Hanger Lane*, Ealing, west London, led to Hanger Hill, itself so named for its 'hanging' wood. *Hempshaw Lane*, Stockport (GtM), led to a copse of saplings, from Old English *impa*, 'sapling' (modern *imp*, meaning first 'offspring' then 'mischievous child') and *sceaga*, 'copse'. *Hills Road*, Cambridge, is certainly not hilly. It may thus be supposed to derive from a person bearing this surname. It is actually so named for its destination, the Gog Magog Hills. *Leechwell Street*, Totnes (Dev), is another spring name, as the street led originally to the 'spring where leeches were found'. A misleading directional name is that of *Linen Street*, Warwick. Its earlier record as *Levenhull Lane* shows that it does not relate to cloth or flax. It led to Levenhull ('Lēofa's hill'), a hill now in the middle of Lammas Field Golf Course.

Lode Lane, Solihull (WMd), led to Lode Heath, while *Minster Moorgate*, Beverley (Hum), was originally a street that ran past Beverley Minster to the moor.

Occupation Road is found in Southwark, south London, and elsewhere. It originated as a name for a lane that led to arable strips in open fields owned by a villager who had *occupation* rights there, as a person who was guaranteed access to his land along the lane itself.

Parliament Hill, Hampstead, northwest London, leads to the hill of the same name that now forms the eastern part of Hampstead Heath. The origin of the hill's own name remains uncertain, despite much speculation to the contrary. Similarly, *Port Hill*, Hertford, leads to the identically named hill ('town hill'), while *Tothill Road*, Plymouth (Dev), leads to Tothill ('lookout hill'). *Lower* (and *Upper*) *Warrengate*, Wakefield (WYo), originally led to a park where there was a warren. *Wood Street*, Barnet, north London, led to a wood, as did (or still do) many other identically named roads, streets or lanes. (But the name has other possibilities equally: the name could refer to a place where wood was obtainable, or derive from someone called Wood.)

The fact that *-gate* can mean 'gate' or 'street' has problems of ambiguity for a common name such as *Westgate*, as mentioned in Chapter 1. Gloucester's *Westgate Street* thus led to the city's western gate. In Peterborough (Cam) and Mansfield (Not), however, as well as in other northern and eastern towns, *Westgate* is simply the equivalent of 'West Street'. West London's *Western Avenue* is more straightforward, with a self-descriptive name similar to that of the *Great West Road*. It was built at the same time as the latter (from the 1920s) and extends under this name from Hammersmith to Denham (Buc), where it leads into the A40 and M40 to Oxford. The A40 becomes the *Oxford Road* at this point on its way westward through Gerrards Cross (Buc).

Street names ending in *-gate* are most common for buildings and other manmade structures and objects. Lincoln's *Bailgate* leads to the old city gate of this name (at the entrance to *The Bail*, or outer court of the castle). *Bishopgate*, Norwich, led to the city gate at the western end of *Bishop Bridge*. *Botchergate*, Carlisle (Cum), has a more complex name. It was originally the name of a gate ('Bochard's gate'), but then under the influence of Old Norse *gata*, 'street', became the name of the street leading to it. Both *Briggate*, Leeds (WYo), and *Bridge Gate*, Derby, have identical names, indicating a street leading to a bridge. Derby's street was itself originally *Briggate*. Elsewhere, *Bridge Street* is commonly found for such a street in many towns and cities, including Abingdon (Oxf), Andover (Ham), Godalming (Sur), Hitchin (Her), Leatherhead (Sur), Maidenhead (Ber), Newbury (Ber), Northampton, Peterborough (Cam), Salisbury (Wil), Stratford-on-Avon (War), Taunton (Som) and Witney (Oxf). Knaresborough (NYo) has a *Briggate* corresponding exactly to its Leeds namesake.

Castlegate may also ambiguously name the gate of a castle or the street leading to it. In York, *Castlegate* indeed led to the castle, as *Castle Gate* did in Newark (Not) and Nottingham. *Derngate*, Northampton, led to the town gate of this name, itself deriving from Old English *derne*, 'secret', although the sense of this is rather obscure. It may have been a gate that was hidden from view. It has its counterpart in *Durngate*, Dorchester (Dor) and Winchester (Ham). *Fengates Road*, Redhill (Sur), led to a gate by a fen or marsh. Both Newcastle (T&W) and Richmond (NYo) have a *Gallowgate*, a street that led to a gallows. Leicester's *Gallowtree Gate* did likewise, with *gallow-tree* actually being a term for a gallows.

Kirkgate is a common north of England name for a street that leads to a church (see Chapter 6, p. 73). *Lairgate*, Beverley (Hum), however, originally led to a barn (Old Norse *hlatha*). *Mill Gate* led to a mill in Newark (Not), as its name implies. *Mountergate*, Norwich, was originally *Parmentergate*, 'street of the furriers', from Old French *permentier*. It did not thus lead to a 'mount', as it seems to suggest. The full form of the name is preserved in that of the church of St Peter *Permountergate*, west of King Street. The familiar *Notting Hill Gate*, west London, takes its name from the turnpike gate that at one time stood here on Notting Hill. *Queen's Gate* in Westminster, west London, leads to the gate of the same name that gives entry to Kensington Gardens. (The queen is Victoria.) *Rickergate*, Carlisle (Cum), is a street that originally led to 'Richard's gate'. *Spitalgate Lane*, Cirencester (Glo), led to Spital Gate, itself named for St John's Hospital here. *Watergate*, Chester, led to the water-gate, that is, to the gate on the northwest wall of the city where the tidal water of the River Dee flowed past. This name is not quite the same as that of *Watergate Lane*, Lewes (ESx), which led to a nearby mill, itself so called for the sluice on its main watercourse. In some instances the word *gate* is missing from a street name of this kind, or is simply implicit. One example is *Lawford Street*, Bristol, which leads to *Lawfords Gate* ('the lord's gate'). Others are *High Street East* and *High Street West*, Dorchester (Dor), which originally led respectively to the town's eastern gate and western gate. The two streets together now form the main east-west road (the A35) through the town.

The generic term *Approach* is used virtually exclusively of modern streets leading to buildings or manmade structures, especially those serving the public. Apart from the ubiquitous *Station Approach* already quoted, examples include *Concert Hall Approach* in Southwark, south London, leading to the Royal Festival Hall on the South Bank, *Post Office Approach*, Forest Gate, east London, leading to the post office at the junction of Romford Road and Woodgrange Road, *Railway Approach*, Southwark, providing access to London Bridge Station (the name is also found in many other towns and districts) and *Tower Bridge Approach*, Stepney, east London, as the northern way past the Tower of London to the famous bridge. (A reader of *The Times* suggested in a letter to the editor of 4 June 1990 that the latter be renamed *Queen Elizabeth Way* in honour of the 90th birthday of Queen Elizabeth the Queen Mother. To date, however, the name remains unchanged. In the 18th century the same road was known as *Iron Bridge*.)

Streets and roads leading to bridges have a variety of names. *Cowbridge*, Hertford, leads to the old bridge over the River Beane; *Fye Bridge Street*, Norwich, leads to Fye Bridge, the principal crossing over the River Wensum; *Hythe Bridge Street*, Oxford, leads to the bridge of this name over the Thames; *Knightsbridge*, Westminster, west London, led originally to a bridge over the Westbourne. The word *bridge* does not always appear in such a name. *Prince of Wales Road*, Dorchester (Dor), leads to Prince's Bridge over the River Frome. For the origins of bridge names themselves, including those mentioned here, see Chapter 10, p. 132.

Streets leading to a ferry crossing rather than a bridge include *Ferry Street*, Lambeth, south London and *Horseferry Road*, Westminster, west London, both giving access to the Thames. *Heath Street*, Barking, east London, is a misleading name, referring not to a heath but to a *hythe* or landing place on the River Roding. Old records of the name confirm this origin. Names also in this 'riverside' category are those that refer to a quay. A common example is *Quay Street*, as in Bristol, Gloucester and Worcester. *Southside Street*, Plymouth (Dev), was recorded as *Southeside Kaye* in 1584, referring to a former quay or wharf by Sutton Harbour. *New Crane Street*, Chester, led in the 18th century to a crane on a wharf by the River Dee. *Pier Road*, Erith, east London, leads to Railway Pier by the Thames.

A road leading to a river is frequently named not for a bridge or ferry but for the river itself. In some cases the road is identically named on both sides of a bridge, but more usually on one side only. Nor does it necessarily follow that a street bearing the name of a river leads to a crossing place. The road may simply follow the course of the river and so be named for it for that reason.

Streets named for the river to which they lead or led include the following: *Aire Street*, Leeds (WYo), *Anton Road*, Andover (Ham), *Arun Street*, Arundel (WSx), *Avon Street*, Bath (Avn), *Chelmer Road*, Chelmsford (Ess), *Colne Bank Avenue*, Colchester (Ess), *Dee Lane*, Chester, *Derwent Street*, Derby,

Douglas Road, Wigan (GtM), *Dun Street*, Sheffield (SYo) (to the Don), *Exe Street*, Exeter (Dev), *Fleet Street*, City of London, *Fossgate*, York, *Irwell Street*, Salford (GtM), *Kent Street*, Kendal (Cum), *Lune Road*, Lancaster, *Medina Road*, Cowes (IoW), *Medway Street*, Maidstone (Ken), *Mersey Street*, Warrington (Che), *Upper Orwell Street* and *Lower Orwell Street* (as a continuous street), Ipswich (Suf), *Ousegate*, York, *Severn Street*, Shrewsbury (Shr), *Severn Street*, Worcester, *High Skellgate* and *Low Skellgate* (a continuous street), Ripon (NYo), *Soar Lane*, Leicester, *Stort Road*, Bishop's Stortford (Her), *Stour Road*, Christchurch (Dor), *Tamedrive*, Tamworth (Sta), *Taw Vale*, Barnstaple (Dev), *Teign Road*, Newton Abbot (Dev), *Thames Street*, Oxford (although now extended to link St Aldate's with Oxpens Road), *Thames Street*, Windsor (Ber), *Wandle Way*, Wandsworth, southwest London, *Weaver Way*, Northwich (Che), *Wensum Street*, Norwich, *Wheelock Street*, Middlewich (Che), *Witham Road*, Grantham (Lin) and *Witham Street*, Boston (Lin). The streets in Chelmsford, Christchurch, Colchester, Derby and Tamworth bear the name on both sides of the river. *Weaver Way*, Northwich, both leads to the river and runs beside it.

By contrast, thoroughfares named for a river because they run alongside it (so that they are *not* directional names) include the following: *Bourne Avenue*, Bournemouth (Dor), *Cherwell Street*, Banbury (Oxf), *Darenth Road*, Dartford (Ken) (beside the Darent, but also leading to the village of Darenth, southeast of Dartford, the village itself being named for the river), *Dour Street*, Dover (Ken), *Humber Street*, Hull (Hum), *Lune Street*, Lancaster (compare *Lune Road*, above), *Medina Avenue*, Newport (IoW) (compare *Medina Road*, above), *Medway Street*, Chatham (Ken), *Neckinger*, Bermondsey, south London (for the stream that ran north into the Thames here), *Nene Way*, Peterborough (Cam) (a lengthy footpath along the north bank), *Ock Street*, Abingdon (Oxf), *Severn Road*, Gloucester, *Sheaf Street*, Sheffield (SYo), *Stert Street*, Abingdon (Oxf), *Tarrant Street*, Arundel (WSx) (from the old name of the present River

Arun), *Thames Street*, Kingston upon Thames, southwest London, *Tib Street*, Manchester, *Toneway*, Taunton (Som), *Tweed Street*, Berwick (Ntb), *Wallington Way*, Fareham (Ham), *Worth Way*, Keighley (WYo) and *Wye Street*, Hereford. *Bridgewater Street*, Sale (GtM), runs by Bridgewater Canal. In the City of London *Upper Thames Street* and *Lower Thames Street* run along the north bank of the Thames. Both streets were originally known together as *Thames Street*, a name dating from at least as early as the 13th century.

The river that for any reason gave its name to a street may now have long been underground, and even channelled as a sewer. This is the case with London's *Fleet Street* and Manchester's *Tib Street*, for example.

In some towns a street named for the local river neither leads to the river nor runs beside it. This is the case, for example, with *Irwell Street*, Bury (GtM), which although running parallel to the river is some distance east of it.

When the name is general, such as *River Street*, *Brook Street* or *Water Street*, only an examination of the local topography will determine whether the street leads (or led) to the watercourse or runs (ran) beside it. *River Street*, Clerkenwell, north London, follows the former head of the manmade New River. *Brook Street*, in Mayfair, west central London, traces the course of the old Tyburn Brook. *Water Street*, Stamford (Lin), runs along the southern side of the Welland. In the 13th century it was known as *Estebythewater* and led to a water mill. On the other hand, *River Lane*, Norwich, formerly ran down to the Wensum, *Water Lane* and *Water Street*, Sheffield (SYo), both approach the south bank of the River Don, and *Water Street*, Westminster, west London, originally descended from Maltravers Street to what was the waterside of the Thames before the construction of the Victoria Embankment. *Water Furlong*, Stamford, runs down to riverside meadows by the Welland. The second word of its name refers to its approximate length (220 yards, or just over 200 metres).

Water Lane is frequently a directional name, even when the way is merely a footpath. Apart from the instance just cited, the road of the name in Berkhamsted (Her) leads to the Bulbourne (and coincidentally the more recent Grand Union Canal), in Norwich to the Wensum (compare *River Lane* in the same city, above), in Salisbury (Wil) to the Avon, in Watford (Her) to the Colne, in Westminster, west London, from the Strand to the Thames, and in York to the Ouse. On the other hand in Winchester (Ham) *Water Lane* runs by the Itchen, not to it.

Two Waters Road, Hemel Hempstead (Her), runs south from the town centre to the Rivers Gade and Bulbourne. It directly takes its name from the district of Two Waters here, however, itself named for the meeting place of the two rivers as early as the 13th century.

Many streets lead or formerly led to a park, and are named for it. Examples are the *Park Street* found in Chester, Cirencester (Glo) and Guildford (Sur), among other places, although the park no longer exists in Chester or Guildford. Similar names are those of *Little Park Street* and *Much Park Street*, Coventry (WMd), both named for the former park here which belonged to the monks and which was divided in medieval times into the Much ('Great') Park and the Little Park. In Lincoln, *Arboretum Avenue* leads to the park of this name (a general term for a botanical garden of trees), while in Highgate, north London, *Waterlow Road* leads to the park that was itself so named because it formed part of the estate of Sir Sydney Waterlow, Lord Mayor of London in the early 1870s.

Other streets lead to a burial place or cemetery. Such names include *Dead Lane*, Loughborough (Lei), *Lich Street*, Worcester (from Old English *līc*, 'corpse', which gave modern *lychgate*; the street leads to the cemetery by Worcester Cathedral) and *Litchdon Street*, Barnstaple (Dev) (from Old English *līc-tūn*, literally 'corpse enclosure', ie cemetery). *Bridle Road*, Maidenhead (Ber), has a name that is said to be a corruption (or euphemistic alteration) of *Burial Road*. *Bury Street*, Guildford (Sur), on the other hand, takes its name from the historic manor here (Old English *burh*). Some towns have streets named for a former gallows or gibbet, such as *Gallows Street*, Warwick and *Gibbet Street*, Halifax (WYo). The gibbet on Halifax Moor was notorious.

Individual buildings frequently give their names to the streets that lead to them. Among them are *Castle Drive*, Chester, *Castle Street*, Salisbury (Wil) (leading to the former castle at Old Sarum, north of the present city), *College Street*, Gloucester (actually leading to the abbey), *Commonhall Street*, Chester (to the old public hall), *Great College Street*, Westminster, west London (to Westminster School, originally St Peter's College), *Guards Club Road*, Maidenhead (Ber) (to the former clubhouse of the Brigade of Guards), *Portobello Road*, Kensington, south London (to Porto Bello Farm, demolished in about 1870) and *Sainsbury Way*, Hull (Hum) (to Sainsbury's supermarket, opened in 1983).

Other directional names of various kinds include *Arcade Street*, Norwich, running from Castle Meadow to the Arcade, *Artillery Row*, Westminster, west London, leading to a former artillery practice ground (ie for shooting with crossbows and handguns), *Beacon Road*, Loughborough (Lei), as a way to a historic hilltop beacon in Charnwood Forest, *Entry Lane*, Kendal (Cum), giving access to one of the town's many 'yards', *Fishpool Street*, St Albans (Her), leading to the abbey fish ponds, *Football Lane*, Harrow, west London, running to Harrow School playing fields, *Holywell Street*, Chesterfield (Der) and elsewhere, leading to a holy spring, *Laystall Street*, Islington, north London, as a passage to a *laystall* (a place where people laid their refuse: see *Mount Pleasant* in Chapter 12, p. 176), *Portmill Lane*, Hitchin (Her), to the mill of this name (meaning 'town mill'), *Pound Lane*, Dorchester (Dor), as a way to a *pound* (an enclosure for stray cattle), *Quarry Street*, Guildford (Sur), to the quarry in Pewley Hill, south of the city, *Standhill Road*, Hitchin, also to a quarry (literally to 'stone diggings', from Old English *stān gedelf*) and *Tennis Court Road*, Cambridge, to a court where real tennis was played (the name dates from at least the 16th century).

As always, some street names can outwardly conceal their origin, even when relatively modern. *Great Central Street*, Marylebone, northwest London, is the short approach street from Marylebone Road to Marylebone railway station, formerly the terminus of the Great Central Line. Similar is *Great Western Road*, Dorchester (Dor), which runs to the town's station, at one time owned by the Great Western Railway. Much older, but equally misleading, is the name of *Tottenham Court Road*, Bloomsbury, north central London. This originally led to the manor of *Tottenhale*. The latter's name became confused with that of Tottenham, north London, however, so that the street bears the altered name accordingly.

6. RELIGIOUS NAMES

There can hardly be a town in the country without a small but distinct proportion of street names of religious origin, if only because a parish church will have its access route, as street or footpath (or both), and because this route will usually be named for it.

Many religious names date back to medieval times, so that the older the town, the greater their proportion. They are broadly of two types: for people or for places. 'People' includes the clergy and members of religious communities; 'places' includes religious houses and places of worship. The latter category also involves names derived from parts of buildings and from religious structures, such as an abbey gate or a wayside cross.

Names of general ecclesiastical reference are fairly common. *Abbots Lane* in Bermondsey, south London, refers to the former Bermondsey Abbey here near which its superior, the Abbot of Battle, near Hastings (ESx), would lodge when paying a visit. *Bishop Street*, Coventry (WMd), was the route by which the bishop would proceed to the priory church of St Michael. *Cannon Row*, Westminster, west London, as the 16th-century antiquarian John Stow reminds us, was 'so called for that the same belonged to the Deane and Chanons (*sic*) of St Stephens chappell who were there lodged'. (St Stephen's Chapel was the royal chapel of the Palace of Westminster.) The same ecclesiastical rank lies behind the name of *Canbury Park Road*, Kingston upon Thames, southwest London, so called from *Canbury*, the manor (Old English *burh*) held here by the canons of Merton Priory. *Dean's Yard*, also in Westminster, was where the Deans of Westminster lived. *Friars Lane*, Lincoln, led to the Dominican house of preaching friars, while *Friar's Entry*, Oxford, is said to be named for the Carmelite friars who lived nearby, and who would have used the lane on their route to the south chapel of the church of St Mary Magdalen. *Friar Street*, Worcester, takes its name from a house of Franciscan friars here. *Friern Road*, Dulwich, southeast London, preserves the name of the former manor of *Freryn Camerwell*, representing land granted to Holywell Priory, Middlesex. *Friern* evolved from *freren*, the plural of Middle English *frere*, 'brother', 'friar'.

Monks Road, Lincoln, takes its name from the nearby Monks Leys, meadows belonging to the Benedictine monks at the priory here. *Monkspath Street*, Solihull (WMd), is apparently named for the path used by the monks at Bordesley Abbey when they made their way to their daughter house at Merevale. *Nuns Road*, Chester, passes over the site of the former St Mary's nunnery. It is not often realised that London's famous *Mincing Lane*, by the Tower of London in the City, is also named for nuns. A convent of nuns or *minchens* owned property here. (*Minchen* is in origin a feminine form of *monk*.) The same word lies behind the name of *Minchenden Crescent*, Southgate, north London, where land was held by the nuns of Clerkenwell. *Prebend Street* is one of a group of ecclesiastical dignitary names by St Paul's Cathedral, London. Others are *Canon Street* and *Rector Street. Priest Lane*, Ripon (NYo), was on manorial land belonging to the canons of Ripon Minster (Cathedral now). *Priestlands*, Sherborne (Dor), is said to be so named from the glebe land granted in 1177 to the priest of St Thomas's Chapel. *Preston Street*, Exeter (Dev), was similarly a street associated with priests. The widespread place-name *Preston* itself means 'priests' village'. It does not necessarily mean that priests lived there, but that the place served as an endowment for priests working elsewhere.

Some names relate to specific religious orders. *Austin Friars*, in the City of London, takes its name from the Austin (ie Augustinian) friars who were given land here in the mid-13th century on which to build a monastery. *Austin Street* and *Austin Friars Lane*, Stamford (Lin), are so named for the former Augustinian friary that was built nearby in 1342 on land now between Rutland Terrace and the River Welland. *Blackfriars*, Gloucester, is named from the monastic house of Dominicans or Black Friars (so named from the colour of their robes) founded here in about 1239. *Blackfriars Lane* in the City of London is named for the monastery of the same order that was built here in about 1276. The identically named street in Stamford preserves a reference to the Dominican monastery that was founded on the site here between modern St Leonard's Street and the Welland at about the same time as the Augustinian house already mentioned.

Carmelite Street in the City of London, together with its continuation, *Whitefriars Street*, recall the Carmelite house built here in the 13th century. Like the Black Friars (Dominicans), the Carmelite White Friars were so named from the colour of their habit. *Whitefriars Street* in Coventry (WMd) refers to the similar house established here in 1343. *Carthusian Street*, also in the City of London, represents the former monastery of the Carthusian order, founded here in 1371. By a process of folk etymology, a Carthusian monastery came to be known as a *charterhouse*, and this word is preserved for the same foundation in *Charterhouse Square* here. (It was this monastic house that gave the name of Charterhouse School, founded on the site in 1611, but moving to Godalming (Sur) in 1872.) *Cluny Place* in Bermondsey, south London, retains a historic reference to the Cluniac monks who owned and administered Bermondsey Abbey until 1381. *Crouch Street*, Colchester (Ess), takes its name from the monastery that the Crutched Friars founded here in about 1244. They were so called because they were 'crutched' or 'crossed', that is, they wore a cross on their habit as their distinctive token. London's *Crutched Friars* in the City is similarly named. The Friary of the Holy Cross was founded here in nearby Hart Street at the end of the 13th century. The Franciscans or Grey Friars (also from the colour of their robes) likewise have streets that preserve their name. *Greyfriars*, Gloucester, commemorates the Franciscan house founded in about 1231, while in Norwich *Greyfriars Road* marks the site of the house founded by the Franciscan friars who settled here in 1226.

Magpie Alley in the City of London may be a religious name in this category, referring again to the Carmelite friars, who although known chiefly as White Friars were also nicknamed Pied Friars, from the contrasting black and white of their robes. It is possible, however, that this particular street may take its name from an inn (see Chapter 9, p. 129). *The Minories* is likewise in the City of London. The name represents the *Sorores Minores* or Little Sisters, an order of nuns founded by St Clare of Assisi in 1215, who established their convent here in 1293. (The first syllable of the name is pronounced as in *minister*, not as in *minor*.)

It may be convenient to include here *Palmer Lane*, Coventry (WMd), which is believed to take its name from the *palmers* or mendicant friars who passed along it on their way between St John's Hospital and the monastery or priory here. (They were so called from the pilgrims to the Holy Land who returned bearing a palm branch or palm leaf.) Again, this street could equally derive its name from the family of a man called Palmer.

Occasionally a name is found that relates to a whole religious group or sect. Such is *Baptist Gardens*, Kentish Town, north London, referring to a Baptist chapel here. Somewhat similarly, *Independent Row* in Ilkeston (Der) refers to the Independent or Congregationalist chapel built here in 1781. *Wesleyan Place*, also in Kentish Town, likewise relates to a Methodist place of worship. At the other end of the doctrinal scale is *Roman Entry*, Manchester, describing the alley that led to the makeshift chapel off Church Street (appropriately enough) where members of the city's Roman Catholic community worshipped in the first half of the 18th century.

Religious foundations established by various orders of monks and nuns are commemorated in several street names. *Convent Lane* in Hull (Hum) takes its name from the convent of Our Lady of Mercy that was founded here as recently as the 19th century. The street was originally *Mill Lane*, and when a new name was sought for it to avoid confusion with an identically named street nearby, the original proposal was for *Nunnery Lane*. The convent authorities objected, however, so the more specific name was given instead. A convent lies behind the familiar name of *Covent Garden*, in Westminster, west London. The reference is to the large herb and vegetable garden which belonged to the abbey of St Peter here in the 13th century. It is purely a coincidence that a fruit and vegetable market was established on the same site in the late 17th century. At the same time it is very likely that produce from the monastery garden would also have been sold to the public. The street called *Covent Close* in Abingdon (Oxf) is a similar if modern commemoration of the famous Benedictine abbey that had been founded here in the 7th century. (*Convent* was at one time equally used of a community of either monks or nuns.)

Many historic towns have a street named after a former friary. *Friary Lane*, Dorchester (Dor), relates to the Franciscan friary founded here in about 1267. *Friary Street*, Guildford (Sur), preserves the memory of the house of Dominican friars founded here in about 1280. (The name is now more generally familiar from the Friary shopping centre on the site.) In Salisbury (Wil) *Friary Lane* recalls the Franciscans who were granted land here in 1357. A priory may be commemorated historically also. This is the case for the religious houses, no longer existing, referred to in the names of *Priory Street*, Hertford and *Priory Road*, Warwick. Hertford Priory was founded in about 1100 and was originally outside the town. *Priory Road* in Loughborough (Lei) has retained a memory of nearby Ulverscroft Priory, now long ruined, and in Stamford (Lin) the identically named street today runs past the remains (some two thirds of the nave) of St Leonard's Priory, east of the town centre.

Several famous hospitals were originally religious foundations, as attested by their religious names, such as London's well-known St Bartholomew's ('Bart's') and St Thomas's hospitals. In a street name, *hospital* may thus refer to such a foundation. It does so in *Hospital Street*, Nantwich (Che), commemorating the hospital of St Nicholas, founded in the 11th century. A charitable foundation of this kind was also known as a *spittle* or (later) *spital*, so that street names containing this word usually have an identical reference. One example is in Lewes

(ESx), where *Spital Street* ran by the former hospital of St Nicholas. London's *Spital Square*, in Spitalfields, east of the City, is named for the hospital and priory of St Mary, founded here in about 1200 on the site of what is now Spitalfields Market. *Spital Hill*, *Spital Lane* and *Spital Street*, Sheffield (SYo), are all named for the Hospital of St Leonard, a Norman foundation. In Peterborough (Cam) St Leonard's Hospital similarly gave the name of *Spital Bridge* in that city. It was founded as a hospital for lepers, but later became a priory and grange. Another medieval hospital for lepers, that of St Giles, lies behind the name of *Spital Walk*, Chester. The yard of its chapel survives as St Giles's Cemetery.

In some instances a street is named from a school or college that was itself originally a religious foundation or that was named for a religious establishment standing earlier on its site. Examples are *Emmanuel Street*, *Jesus Lane* and *Trinity Lane*, Cambridge. But such names relate to (and derive from) the educational establishment, not the original religious house, so belong properly to Chapter 8, not here. Emmanuel College, for example, was founded only in 1584 on the site of a religious house occupied by Dominican friars that had been dissolved in 1538 but that was originally founded in 1240. *Emmanuel Street* is thus named for the college, and was recorded in 1544 as *Blak Fryers Lane*, for the monks, who had enclosed the lane in the 13th century when they enlarged their cemetery. Jesus College was similarly founded in 1496 on the site of the priory of St Radegund, so that *Jesus Lane*, named for the college, was earlier *Radegundeslane* (so recorded in 1359) and then *Nunneslane* (in 1435). Insofar as these early names are of direct religious origin, however, they deserve their place here, even though their successors do not. *College Street*, Northampton, however, takes its name from the house of the collegiate clergy of All Saints, founded in the 14th century, so has its rightful place in this chapter. Here, too, mention should be made of *Gloucester Green*, Oxford. It takes its name from Gloucester Hall, a Benedictine monastery founded in 1283 on the site of the present Worcester

College. The house's own name derives from its connection with Gloucester Abbey.

Monasteries, priories and convents may be regarded as 'private' establishments, intended for the religious community to whom they belong and who originally founded them. Abbeys, churches, cathedrals and chapels, on the other hand, although in many cases also founded by religious communities, may be thought of as 'public' places of worship, open to all. It follows that streets named for public places of worship have the most common religious names, with *Church Street* (or *Church Road*, or *Church Lane*) found in dozens of towns and villages (see the Appendix, p. 194). Two instances of many are *Church Street*, Stratford-on-Avon (War), leading to Holy Trinity Church (and so named since at least the early 14th century) and *Church Street*, Warwick, from St Mary's Church, but recorded no earlier than 1516. London's *Kensington Church Street*, between Notting Hill Gate and Kensington High Street, leads to St Mary Abbot's church. It was earlier simply *Church Lane* (thus in 1658) and as late as 1851 appears on a map of London as *Chirch Lane. Kensington* was subsequently prefixed to distinguish this particular Church Street from the many others.

Chapel Street is almost equally ubiquitous. Stratford-on-Avon (War), just instanced, has not only *Church Street* but both *Chapel Street* and *Chapel Lane*, each referring to the Guild Chapel and each first recorded in the 15th century. *Chapel Street*, Petersfield (Ham), is named for the former St Andrew's chapel which stood at its northern end. More originally, Norwich has *Chapel Loke* as the name of a narrow lane between Surrey Street and Ber Street. It leads to the nondenominational Surrey Chapel, with *loke* an East Anglian dialect word meaning 'lane' (from Old English *loc*, 'fold', 'enclosure', modern *lock*). *Chapelhay Street*, Weymouth (Dor), was originally simply *Chapel Street* (so recorded in 1824). It was then renamed from the *Chapel Hay*, the field or enclosure by the historic chapel of St Nicholas, which was converted into a fort in the Civil War.

Elsewhere may be found *Abbey Street* in Bermondsey, south London, preserving the memory of Bermondsey Abbey, founded here in 1082, *Cathedral Street*, Southwark, south London, beside the cathedral of St Saviour, *Minster Street*, Salisbury (Wil), leading to the 'minster' (cathedral), and originally extending along the route now covered by Castle Street, Minster Street and High Street (the name dates from the 13th century), *Minster Avenue*, Beverley (Hum), for Beverley Minster, originally founded as a monastery in the 8th century (compare *Minster Moorgate* in Chapter 5, p. 64, above) and *Tabernacle Street*, Shoreditch, northeast London, so named for the Methodist place of worship established here in the 1740s.

A *beadhouse* was a term formerly used for a place of worship, from Old English *bed-hūs*, literally 'prayer house'. It lies behind the name of *Bede House Lane*, Newark (Not). An Old English term of almost identical meaning was *bed-ærn* (literally 'prayer building'), which in turn gave the names of both *Bedern Bank*, Ripon (NYo) and *The Bedern*, York. Ripon's prayer house originated as a residence for six vicars in the early 14th century, while in York the 'prayer building' was the residence of the vicars choral of the minster, its name dating from 1275, when it was recorded exactly as now. By contrast, *Bethel Street*, Norwich is so called not from *Bethel* as a name (of biblical origin) for a Methodist or Baptist church but from the *Bethel Hospital* or *Bedlam Hospital* opened here as a mental institution in the early 18th century. *Zoar Street*, however, in Southwark, south London, is named for the Baptist Zoar Chapel. The biblical city of Zoar escaped destruction because Lot took refuge there. As a result, its name was adopted for Nonconformist chapels regarded as places of sanctuary.

Streets named for a particular church's dedication are very common. Most obvious are those derived from a saint's name preceded by *St*. Examples include the following: *St Ann's Square*, Manchester (for St *Ann*'s church, built in 1709 by Lady *Ann* Bland in the reign of Queen *Anne*), *St Barnabas Street*, Westminster, west London (for the church built here in 1848), *St*

Giles', Oxford (from the church built here in the early 13th century), *St John Maddermarket*, Norwich, formerly *St John's Street* (for the church of St John the Baptist founded before the Conquest and named for the market where *madder* was sold in order to be distinguished from the church of St John the Evangelist), *St Leonard's Street*, Stamford (Lin) (leading to the priory of St Leonard, founded in the late 11th century: compare *Priory Road*, above), *St Martin's Lane*, Westminster, London (for a church of the name that has stood here since the 11th century), *St Michael's Hill*, Bristol (for a church dating from the late 12th century, if not earlier), *St Nicholas Church Street*, Warwick, formerly *St Nicholas Street* (from the old church here), *St Stephen's Square*, Hull (Hum) (for the Victorian church built here in 1844 but demolished in 1955), *St Swithin's Lane*, City of London (for the church, rebuilt in the 15th century, that was bombed here in the Second World War).

Oxford's *St Aldate's*, earlier known as *Fish Street*, has a rare name, for a church with a dedication known only here and at Gloucester. The church is first recorded in the early 12th century, and its name is believed to be a corruption of *old gate*, so that St Aldate himself never actually existed. Both this church and the one in Gloucester are near the site of old gates.

Some streets apparently named after a church's sanctual dedication are actually transferred names, adopted from elsewhere. One example is *St Leonards Road*, Acton, west London, named for the village near Wendover (Buc) where the owners of the land here had property. *St Leonards Terrace*, Chelsea, west London, takes its name from the home village of the man who built the street, Upton St Leonards near Cheltenham (Glo).

Streets may be named after a church dedication of this kind but without the prefix *St*, which may obscure the actual origin.

Thus *Benet Street*, Cambridge (more pedantically spelt *Bene't Street*), takes its name from the Saxon church of St Bene't (ie St Benedict), and in the same city *Botolph Lane* is named for the medieval hospice of St Botolph.

Catherine Street, Plymouth (Dev), is so named for the former chapel dedicated to St Catherine. *Catherine Street*, Salisbury (Wil), however, has a name that is a corruption, and belongs to Chapter 7, not here (see p. 98).

Cecily Hill, Cirencester (Glo), is named from the chapel of St Cecilia. *Clementhorpe*, York, is so called from the Benedictine nunnery and church here dedicated to St Clement, with *thorp* meaning 'hamlet', 'settlement'. The name dates from the 11th century. *Foster Lane*, in the City of London, bears a name that is a medieval corruption of that of St Vedast, a 6th-century Flemish saint. The same saint gave the name of *St Faith's Lane*, Norwich, now *Mountergate* (see Chapter 5, p. 65). The unfamiliar saint's name became confused with that of the better-known St Faith the Virgin. It is preserved in Norwich, however, in the relatively modern *St Vedast Street*, earlier *South Cathedral Street*.

John Street, Bristol, is named for the medieval church of St John, while *Lawrence Street*, York, takes its name similarly from the church of St Lawrence without Walmgate. *Litchurch Street*, Derby, looks as if it has a name meaning 'little church'. The true origin, however, appears to be in a saint's name (or simply Anglo-Saxon personal name) *Luda*, with whom a church was associated here. Less ambiguously, *Magdalen Street*, in Colchester (Ess), Exeter (Dev) and elsewhere, is named from a church or other religious foundation dedicated to St Mary Magdalene. *Magdalen Street* in Southwark, south London, though, is named for Magdalen College, Oxford, whose founder, William of Waynflete, Bishop of Winchester, lived at Winchester House, Southwark and endowed the college with property here in London. *Maudlin Street*, Bristol is named for a medieval hospice dedicated to the same saint. *Nicholas Street*, Chester, at one time *Blackfriars Lane*, is named for the Dominican church dedicated to this saint. In like manner *Paul Street*, Exeter, takes its name from the church of St Paul. *Peter's Hill* in the City of London is all that is left of a lane that led from St Paul's churchyard to Thames Street. It preserves the memory of the former small church of St Peter, destroyed in the Great Fire of London (1666). *Great Peter Street* in the City of Westminster, on the other hand, was built on land owned by Westminster Abbey and is named for the famous church's dedicatory saint. The name was recorded for the original shorter street here as *Sant Peterstreet* in 1624 and *Peter Street* in 1747. The present name was subsequently applied to *Peter Street* and to its two eastward extensions of *Little Peter Street* and *Wood Street*, making a single street from Horseferry Road to Millbank.

Bristol's *John Street*, already mentioned, has its companions in *Peter Street* and *Philip Street*, named for churches respectively dedicated to these saints. *Powells Walk*, Chiswick, west London, conceals the name of St Paul's Cathedral, whose Dean and Chapter owned the manor here. Rather unexpectedly, *Sidwell Street*, Exeter (Dev), is also named for a saint. She was the virgin martyr Sidwell, said to have been buried near the city. She is commemorated in the church of St Sidwell here. Her own name is perhaps a combination of Old English *sidu*, 'morality' and *full*, so that she was 'entirely virtuous'. Her suspiciously appropriate name, however, gives grounds for doubting her actual existence. Nevertheless the street still leads to the church that is dedicated to her.

London's *Tooley Street*, in Southwark, south of the Thames, preserves the name of St Olave, to whom the church here was dedicated. The final *t* of *saint* became attached to the personal name, as it did for *St Audrey* to produce the modern word *tawdry*. St Olave (Olaf) was the 11th-century king of Norway who fought for England against the Danes.

Church dedications not referring to individual saints are more patently seen in other street names. *All Hallows Lane*, Kendal (Cum), is named for the former chapel of ease here. *All Saints Green*, Norwich, is named for the church of All Saints Timberhill (as distinct from All Saints Fyebridge). On the other hand *All Souls Avenue*, Willesden, northwest London, is named for All Souls College, Oxford, which holds land here. The first of the three names cited here gave that of *Allhalland Street*,

Bideford (Dev), which was known as *Allhallow Street* in the late 17th century. *Trinity Street*, like *All Souls Street* mentioned here, is in some cases named for Trinity College (Oxford or Cambridge), if the college holds land locally. In the case of *Trinity Road*, Wandsworth, southwest London, however, the reference is to the church of the Holy Trinity, Upper Tooting, built in the mid-19th century. The name of *Bow Street*, Durham, gives only part of the dedication from which it derives, that of the church of St Mary-le-Bow.

Similar church dedications can give street names ending in *gate*, often in the sense 'street' but also meaning 'gate'. In the former case the street usually leads (or led) to the church; in the latter the gateway referred to gave access to the religious foundation or its precincts. In Ripon (NYo) *Allhallowgate*, *St Agnesgate* and *St Marygate* are all streets that lead or led to the church of All Hallows, the hospital of St Anne (originally St Agnes) and a chapel dedicated to the Virgin Mary on the site of the former abbey of Ripon. *Gillygate*, in both Pontefract (WYo) and York itself, is a street named for a church dedicated to St Giles. *Ladygate*, Beverley (Hum), is a street leading to St Mary's Church, while *Michaelgate*, Lincoln, near St Michael's church, is a late name, first recorded only in 1842. It may have been influenced by the much older *Micklegate*, the former name of the High Street. York also has *Marygate* and *Petergate*, named respectively for St Mary's Abbey and York Minster, the latter being dedicated to St Peter.

Gate names derived from religious terms of more general origin, as considered earlier, include *Abbeygate*, Leicester, *Blackfriargate*, Hull (Hum), *Deangate*, York, *Friargate*, Derby, *Greyfriar Gate*, Nottingham, *Monkgate*, York, *Priestgate*, Peterborough (Cam) and *Whitefriargate*, Hull. The sense of *gate* in all these is 'street'. For *Temple Gate*, Bristol, however, it is 'gate'. This and *Temple Street* take their names from the church known as *Temple*, founded by the Knights Templars. The same name lies behind that of the more generally familiar Temple Meads railway station here.

Most such names are medieval, but *Deangate*, running along the south side of York Minster, was created only in the early 20th century, when the Dean and Chapter gave land to York Corporation to provide a road for horse traffic. (The present Dean and Chapter have been campaigning since the early 1980s to close the road, as its incessant motor traffic encroaches on the tranquil environment of the minster and creates a constant background noise disruptive to daily services.)

A regular *gate* name of religious origin is simply *Kirkgate*, a northern equivalent of the southern *Church Street*. It is found in Leeds, Newark (Not), Ripon (NYo) and Wakefield (WYo), among other towns. In Wakefield the name undoubtedly refers to the cathedral, past which it runs. But the street that now bears the name here was initially *Kergate*, from Old Norse *kjarr*, 'marsh'. The street thus led down to marshland by the River Calder, and its present name is a medieval alteration of the original form.

Names can relate to particular functions or parts of religious buildings. *Broad Sanctuary* and *Little Sanctuary*, both in Westminster, west London, refer to the refuge known as St Peter's Sanctuary to which they led, the refuge itself being named for its location at the abbey of St Peter, that is, the Benedictine monastery that became the present Westminster Abbey. The sanctuary granted refugees freedom from arrest by the civil authorities. More obliquely, *Chantry Lane*, Westbury (Wil), takes its name from the historic manor which was endowed here by the *chanter* (precentor) of Salisbury Cathedral. *Steeple Row*, Loughborough (Lei), is named for the nearby All Saints' Church. The church actually has a tower, not a spire, but *steeple* is still properly the word for a church tower of any kind.

The Gallery, Ely (Cam), probably took its name from the *galilee* or porch of Ely Cathedral, although local tradition derives the name from the *gallery* that led from the Bishop's Palace to the west transept of the cathedral. *Bow Lane*, in the City of London, is named for the Church of St Mary-le-Bow, itself so named as it is built on arches, while *Chancel Lane*,

Wilmslow (Che), runs past the chancel end of Wilmslow Church. *Cenacle Close*, Hampstead, northwest London appears to take its name from the adjacent Sunnyfield Convent. A *cenacle* is a religious term for a small room in which supper is taken, and properly applies to the 'upper room' in which the Last Supper was held by Christ and his apostles, from Latin *cena*, 'supper'. *The Cenacle*, a modern street in Cambridge, may relate similarly to local ecclesiastical property.

Several towns have streets named for a cross, although *Cross Street* could simply be a street that leads across, or one that runs to a crossroads. London's *St Cross Street*, originally *Cross Street*, is actually of the latter type, despite its religious prefix. It is so named as it crossed the land belonging to the Hatton family here in Holborn, and was renamed so as to be differentiated from the many other *Cross Streets*. Names of genuinely religious origin referring to a cross are *High Cross Street* and *Red Cross Street*, both in Leicester, the latter named for its colour, and *White Cross Street* in the City of London, which led to a white cross. Another word for a cross or crucifix is *rood*, seen in *Holyrood Street*, Newport (IoW) and *Rood Lane*, Coventry (WMd). More obviously self-explanatory is *Crucifix Lane*, Bermondsey, south London. The name refers to the crucifix or cross which was removed from Bermondsey Abbey after its dissolution in about 1539 and set up at the end of this street. In its original siting, the abbey crucifix had become a favourite place of pilgrimage, and had given the earlier name of *Holyrood Street* here. Both streets are almost contiguous, in a location now close to London Bridge railway station.

Holy wells may be commemorated in street names. In Chesterfield (Der) *Holywell Street* is named for a holy well at the former chapel of St Helen. In Oxford the identically named street is a reminder of a well to the north of St Cross Church. The well here is no longer visible, although one may still be seen in St Albans (Her), where it gave its name to *Holywell Hill*, the main road running south past the east end of the cathedral to the River Ver. In Coventry (WMd) *Jordan Well* is a street name perhaps relating to a holy well named for the biblical river Jordan.

Near St Paul's Cathedral in the City of London are some short streets with patently religious names: *Amen Corner*, *Ave Maria Lane*, *Creed Lane* and *Paternoster Row*. These may commemorate craftsmen who made and sold religious objects here, with the first three names relating specifically to texts and prayers. *Paternoster Row* (recorded as *Paternosterstrete* in a document dated 1307) is the oldest of these, and the later names may have been created to match it. A *paternosterer* was a maker of rosaries, the term originating in Latin *pater noster*, 'our father', the opening words of the Lord's Prayer.

In surveying street names of religious origin mention should finally be made of the four streets in Chichester (WSx) called respectively *North Pallant*, *South Pallant*, *East Pallant* and *West Pallant*. The basic name *Pallant* originally applied to the southeast sector of the city, which belonged to the Archbishop of Canterbury rather than to the Bishop of Chichester. It probably represents Late Latin *palantia*, literally 'palace', here denoting an area over which the archbishop had special 'palatine' rights. The name dates from the late 13th century.

7. TRADING, OCCUPATIONAL AND NATIONAL NAMES

The buying and selling of merchandise was a major economic activity in medieval towns. It was carried on either in a market place or in a street that served a dual purpose as both thoroughfare and trading place. The latter has its modern equivalent in the weekly market, held in a street that is usually closed to normal traffic for the purpose.

Most general markets of today, however, are still held in a *Market Place* or *Market Square*, as the central public area of a town affording the greatest space and most suitable accommodation for traders. These names (as well as simply *The Square*) are thus found in many towns. *Market Street* is also not uncommon, as in Guildford (Sur), Manchester, Nottingham and Oxford, among other places. Cambridge has its *Market Hill* as its central market square. There is no hill here now, but local archives show that at one time the area so named stood at a greater height above the River Cam than it does today, and that the general level of the land over to the west of the Cam was gradually raised from the 15th century to the 19th. In medieval times *Market Hill* was known as both *Marketstede* (from Old English *stede*, 'place') and *Marketplace*, and the present name was first recorded in 1520. A similar history applies to *Peas Hill*, also in Cambridge (see below).

Nottingham is not unique in having both a *Market Place* and a *Market Street*. The former is recorded as the *Merketsted* in 1395 and as *The Merke Steede* in 1547. The latter name, however, is of relatively recent date, recorded no earlier than 1866. *Market Street* was previously known as *Sheep Lane*, referring to the type of livestock sold there. Oxford is now best known for its covered market, built in 1772. It likewise has a *Market Street*, known in medieval times as *Cheynelane*, for the chain at its western end, and in the 17th century as *Jesus Colledge Lane*, for the college down one side of which it runs. A third well-known street in the same city is *Cornmarket*, relating to a specific commodity. Its medieval name was *Northyatestreet*, for the north gate into the city here. Other examples of markets where particular merchandise was sold are cited below.

Street names deriving from Old English *cēap*, 'trade', 'merchandise' (modern *cheap*), also refer to a market. The best known examples are in the City of London, where there were two major food markets in medieval times, the eastern and the western. The eastern was *Eastcheap*, and the western at first *Cheap* or *Westcheap*, then *Cheapside*. The precise sense of *side* in this last name is uncertain, but the reference may be to the line of houses that came to be built along one side of the market. Similar names are found elsewhere, such as *Cheap Street* in Newbury (Ber) and Sherborne (Dor), and *Cheapside* in Halifax (WYo) and Nottingham. *Cheapside* in Birmingham and Nottingham are first recorded respectively in 1786 and 1826, and are probably borrowings of the London name. *Wincheap Street* in Canterbury (Ken) is a related name, the first part of which probably represents Old English *wægn*, 'wagon', 'cart' (modern poetic *wain*). In Coventry (WMd) *Cross Cheaping* is a street name relating to a market cross. It dates from the mid-14th century, with the second word representing Old English *cēping*, 'market', familiar as the *Chipping* in the names of towns such as Chipping Campden, Chipping Ongar and Chipping Sodbury.

Even today many of the streets adjoining the London *Cheapside* have names relating to particular trades, including *Wood Street*, *Milk Street*, *Bread Street* and *Ironmonger Lane*. For further examples of such names, see below.

Commercial Road is a modern cousin of *Market Street*. However, the name has the primary sense of 'throughway for commercial traffic' rather than relating to a trading activity, although many streets of the name have subsequently become associated with the businesses and shops that line them, as in Portsmouth (Ham), for example. The name dates from the late 18th or early 19th century. *Commercial Road*, Stepney, east London, was built in 1804 for commercial traffic between the City of London and the newly constructed East India and West

India Docks on the Isle of Dogs. Other towns and cities with streets of the name (in many cases as *Commercial Street*) include Aldershot (Ham), Barnsley (SYo), Barnstaple (Dev), Bedford, Birmingham, Bournemouth (Dor), Bristol, Darlington (Dur), Dewsbury (WYo), Exeter (Dev), Gloucester, Grantham (Lin), Gravesend (Ken) (less commonly *Commercial Place*, by the Customs House), Guildford (Sur), Halifax (WYo), Hanley (Sta), Harrogate (NYo), Hereford, Huddersfield (WYo), Hull (Hum) (formerly *Castle Row* but renamed in 1867), Ipswich (Suf) (linking Stoke Bridge over the Orwell with Station Bridge, built in the mid-19th century), Kettering (Ntp), Leeds (WYo), Lowestoft (Suf), Macclesfield (Che), Manchester, Mansfield (Not) (as *Commercial Gate*, where doubtless the meaning 'street' is intended), Nelson (Lan), Northampton, Plymouth (Dev) (parallel *Commercial Road* and *Commercial Street*), Poole (Dor), Sheffield (SYo), Southampton (Ham), South Shields (T&W), Stockton-on-Tees (Cle), Sunderland (T&W), Swindon (Wil), Weymouth (Dor), Woking (Sur) (as *Commercial Way*) and Wolverhampton (WMd). It will be noticed that many of these are located in either Midland or northern industrial towns or else major ports. The street is often found in a dockland or quayside region, as in Barnstaple, Bristol, Gloucester, Hull, Plymouth and elsewhere. *Commercial Way* in Peckham, south London, linking Southampton Way in the west with the Old Kent Road in the east, bridged the former Surrey Canal. In the 19th century it was known as *Commercial Road*, a name that itself replaced a run of 35 names extending the length of the street. Two of these were *Commercial Place* and *Commercial Terrace*. *Commerce Road* is sometimes found as a related name, for example in Brentford, west London, where it leads to the Thames and Brentford High Street.

Markets named for specific commodities do not necessarily date from medieval times. *Beastmarket Hill*, Nottingham, is first recorded only in 1826, while Oxford's *Cornmarket*, already mentioned, dates from 1772, when it applied to the east side of the street only, the west side retaining the old name of *North*

Gate Street. The actual Cornmarket building in Oxford, however, was built in 1536. (In 1644, at the time of the Civil War, it was pulled down by soldiers so that the leading from the roof could be made into bullets.)

Mention has already been made of the *Maddermarket*, Norwich (see *St John Maddermarket*, Chapter 6, p. 81). The same city's *Cattle Market Street* is first recorded only in 1885. In the 18th century its northern half was known as *Common Pump Street* and its southern half as *Beaumont's Hill*. A livestock market was held in the street from 1740 right up to the 1960s, when it was moved elsewhere. As a former major fishing port, it is hardly surprising that Hastings (ESx) should have its *Fishmarket*, recorded as *Fishmarkett Place* in 1671. London's *Haymarket*, in the City of Westminster, takes its name from the market here where hay was sold as fodder until the 1830s. The same name is found in Bristol. Northampton has its *Horse Market*, dating from 1545.

A regular weekly market can also produce a street name, and Beverley (Hum) has both a *Saturday Market* and a *Wednesday Market*, respectively dating from 1577 and 1467. The town still has a regular cattle market on Wednesday and general market on Saturday, the latter in the named street, the town's principal square. A similar arrangement applied in Nottingham, where the present *Weekday Cross*, in the south of the city, represents a market cross so named by contrast with the former *Saturday Market*, the *Old Market Square* of today. Both these names date from medieval times, the former as *The Wykday Market* in 1336 and the latter as *le Seterday market* in 1353. A name in this category more obviously disguised in modern terms is that of *Market Jew Street*, Penzance (Cor), where the second word represents Cornish *yow*, 'Thursday'. The market here came to be so called by contrast with that at Marazion, three miles away, the name of this smaller town deriving from Cornish *morghas byghan*, 'little market'. Penzance still holds its regular weekly market on Thursdays. Newcastle (T&W) has its

Bigg Market, the name not referring to size but to the grain sold there, from Old Norse *bygg*, 'barley'.

The market name based on *Cheap* has its equivalent specific reference in some towns. One example is *Meal Cheapen Street* in Worcester, recorded in 1369 (as *Melechepyng*), and meaning 'meal market'. The second word of the name is the same as that in Coventry's *Cross Cheaping*, already mentioned. In York, the streets named *Bootham*, *Bootham Row* and *Bootham Terrace* preserve a basic name derived from *búthum*, a West Scandinavian form of an Old Norse name meaning literally 'at the booths'. The abbot and monks of St Mary's Abbey had a right to hold a weekly market here. The name is first recorded (as *Bouthum*) in the mid-12th century.

The most common types of specialised market are those for livestock, grain and dairy products. The first of these produced such names as *Cock Lane* in the City of London, where cockerels were reared for market (or possibly where cock-fighting took place), *Cowgate*, in Peterborough (Cam), where cows were presumably sold in medieval times, perhaps as an 'overflow' from the regular cattle market, *Culver Street*, Salisbury (Wil), where doves or pigeons were kept or sold (from Old English *culfre*, 'dove'), *Goss Street*, Chester, where geese were kept and sold, and *Hengate*, Beverley (Hum), where hens were similarly housed. The name *Horsefair* (or *Horse Fair*) is fairly widely found and is self-explanatory. Towns with such a street include Banbury (Oxf), Bristol, Leicester, Malmesbury (Wil), Pontefract (WYo), Ripon (NYo) and Wetherby (WYo).

Poultry, in the City of London, is a reminder of the site where the poulterers had their stalls, close to Cheapside. The name is found elsewhere, for example as the *Poultry Cross* in Salisbury (Wil), which stands near the site of the former *Poultry Street* (recorded as *Pultri strete* in 1361). This earlier street may have run where modern Silver Street now is. *Rother Street*, Stratford-on-Avon (War), takes its name from the cattle sold here, from Old English *hrīther*, 'ox', 'cattle'. The self-explanatory *Sheep Street* and *Sheepmarket* are found in a number of towns, such as Cheltenham (Glo), Stamford (Lin) and (again) Stratford-on-Avon. Much less obvious is the name of *Stodman Street*, Newark (Not), which probably derives from Middle English *stodmere*, 'stud mare', but which was later confused with *stodman*, 'stud man', that is, a man who worked in a *stud*, where horses were bred. More generally, *Pinfold Street*, Birmingham and elsewhere, takes its name from the word for a *pen* (pound) for stray cattle.

Not all streets with names derived from animals will have related to trading, however. Birmingham's famous *Bull Ring*, for example, was the place where bull-baiting was held, and even *Cowgate* can mean a street along which cows are driven rather than one where they are sold. It probably has the former sense in Norwich. *Goose Gate*, Nottingham, does not mean what it seems to mean, despite the city's historic *Goose Fair*, at which geese were long sold. Early records show that it was named for one Robert *Gos*, a goldsmith, also known by the Norman French form of his name as Robert *le Orfevere*.

As now, animals were sold as meat as well as livestock, and a name that is found in many towns with reference to this is *Shambles*. The word itself represents Old English *sceamol*, 'bench', implying the benches or tables that were set up in the open air for the sale of meat. (In the plural, the word subsequently came to be used for a slaughterhouse; hence modern *shambles* as a scene of disorder or confusion.) The fuller form of the name was *Flesh Shambles*, 'meat benches', as recorded for *The Shambles* in York and Chesterfield (Der), for example. *Newgate Street* in the City of London was also formerly known as *The Shambles*, while the street of this name in Worcester was formerly *Baker Street*. *The Shambles* in Chippenham (Wil), however, is first recorded only in 1570. Kendal (Cum) has both an *Old Shambles* and a *New Shambles*, the latter being a small street between the Market Place and Finkle Street, and originally simply *Shambles* (*Shamyles* in 1452). *Old Shambles* lies west of it, leading off Stricklandgate.

It is only one small step from a shambles to a butchery, and the latter word lies behind such names as *The Butchery*, Ely (Cam), *Butchery Green*, Hertford and *Butchery Lane*, Canterbury (Ken). Not far removed are the names of *Fish Row*, Salisbury (Wil) and *Fish Street*, Hull (Hum). The former city is some distance from the sea, however, whereas Hull has long been a centre of the fishing industry.

Specific commodities such as bread, butter, grain and herbs gave names such as *Bread Street*, City of London, Wakefield (WYo) and elsewhere, *Butter Cross*, Oakham (Lei), *Cinnamon Street*, Stepney, east London, *Cornhill*, City of London, Dorchester (Dor) and Lincoln, the evocative *Land of Green Ginger*, Hull (Hum), perhaps referring to a garden where pot herbs were grown, *Milk Street*, Exeter (Dev), *Oaten Hill*, Canterbury (Ken) and *Oatmeal Row*, Salisbury (Wil), *Peas Hill*, Cambridge (no longer a hill, however, like *Market Hill*, above), where peas were sold, *Pepper Lane*, Coventry (WMd) and *Pepper Street*, Chester, *Rye Street*, Bishop's Stortford (Her), *Westlegate*, Norwich, where *wastel* was sold as bread made from fine white flour, and *Wheat Street*, Nuneaton (War). Here also may be included *Salt Lane*, Salisbury (Wil), where salt was sold, as well as *Salthouse Lane* in Hull and Ipswich (Suf). The streets called *The Maltings* in various towns, such as Kings Langley (Her), Oxted (Sur) and Byfleet (Sur), usually refer to a former malthouse or brewery nearby.

Wood and timber lie behind some street names, such as *Timber Hill*, Norwich, earlier *Tymbermarket hill* (so recorded in 1507), and *Wood Street*, Kingston upon Thames, southwest London and Stratford-on-Avon (War), where wood was stored or sold. *Gorse Stacks*, Chester, takes its name from a nearby common where gorse was stacked as fuel for the city's bakeries. The stacks were deliberately raised outside the city walls because of the risk of fire. Unexpectedly, *Oat Street*, Evesham (H&W), belongs here, as the name has developed from *le Odestreet*, so recorded in the 14th century and itself a form of *Wodestrete* (ie 'wood street'). Evesham also has *Cowl Street*, referring to the coal sold there.

Fabrics and the products made from them have given names such as *Blanket Row*, Hull (Hum), *Cable Street*, Stepney, east London, *Flaxengate*, Lincoln, *Rope Walk*, Lincoln and *Woolshops*, Halifax (WYo), now appropriately absorbed into the shopping centre. *Whimple Street*, Plymouth (Dev) is said to derive its name from the wimples or veils made here in medieval times. The City of London's *Petticoat Lane* belongs here. An old clothes market has been held in the street since at least the 16th century, and is today a colourful Cockney event set up on Sunday mornings. The name, first recorded as *Peticote Lane* in 1602, is now found on many London street plans and is more familiar than the street's official name of *Middlesex Street*. This was given in about 1830 and refers to the fact that the street forms the boundary between the City of London and the historic county of *Middlesex*.

More solid materials and substances, including stones and metals, produced the names of *Brick Lane*, Stepney, east London, where earth was dug to make bricks or tiles, *Clay Street*, Hull (Hum), also so called for its brickmaking industry (although the name was given only in 1958), *Copper Street*, Sheffield (SYo), where this metal was worked or sold, *Coprolite Street*, Ipswich (Suff), from the supposed fossil dung dug locally, *Glasshouse Street*, Nottingham, where glass was made, *Gold Street*, Northampton, from the goldsmiths who traded here, *Lime Street*, City of London, where there were lime kilns, and *Steelhouse Lane*, in Birmingham and Sheffield (SYo), cities famous for their steel industry. In *Crock Street*, Barnstaple (Dev), pots or 'crocks' were made, while *Hart Street*, in the City of London, may have a connection with hearthstones. The latter name was recorded as *Herthstrete* in the mid-14th century, lending support to this origin.

The name *Silver Street* is widely found, for example in Cheltenham (Glo) and no less than eight towns in Wiltshire, among them Calne, Salisbury, Trowbridge and Wilton.

Marlborough (Wil) has *Silverless Street*, which may also originally have been *Silver Street*. *Silver Street* in Lincoln has a relatively modern name, and does not appear to relate to the precious metal. However, a former *Silver Street* is known and was recorded in the mid-14th century. The city also had a *Silver Lane*, believed to be the modern *Mint Lane*, both names referring to the minting of money carried on here. For most streets the name is not medieval in origin but dates from the 16th century or later.

Two market or trading names of general origin are *Newport*, Lincoln and *London Street*, Norwich. The former, in the north of the city, led to the district of the name, itself meaning 'new market' and probably of Norman origin. The latter, in the city centre, doubtless refers to its commercial activity, and is explained by an 18th-century writer as deriving 'possibly from the number of Shops of all Trades & the great traffic and passage of Peoples Horses, Carts, Coaches, &c, through it' (J. Kirkpatrick, *The Streets and Lanes of the City of Norwich*, c.1720).

Several streets are named not so much for the commodity or manufactured article as for the person who trades in it or makes it, that is, not for the bread but for the baker, not for the gold but for the goldsmith.

Most such names relate to specific trades or occupations, but general names are sometimes found, such as *Chipper Lane*, Salisbury (Wil), from Middle English *chipere*, 'market man' (directly related to *Cheapside* mentioned above) and *Merchant Street*, Bristol, first recorded in 1537.

Occupations that are almost all readily understood today include those behind *Baker Street*, Gloucester, *Brewer Street*, Bletchingley (Sur) and Oxford, *Butcher Row*, Coventry (WMd), *Carter Gate*, Newark (Not) and Nottingham, *Colliergate*, York (dealing in coal or charcoal), *Cook Street*, Coventry, *Dyer Street*, Cirencester (Glo) (a town noted for its wool trade), *Fishergate*, Norwich, Nottingham, Ripon (NYo) and elsewhere, also *Fisher Row*, Chesterfield (Der) and Oxford, *Fishers Lane*, Cambridge and *Fisher Street*, Carlisle (Cum), *Goldsmith Street*, Exeter (Dev), *Hosier Lane*, City of London and *Hosier Street*, Reading (Ber) (supplying stockings), *Ironmonger Lane*, City of London, by *Cheapside*, and *Ironmonger Row*, Coventry, *Mercers Row*, Northampton (dealing in silk and velvet), *Pottergate*, Lincoln and Norwich, *Ropemaker Street*, City of London, also *Ropergate*, Pontefract (WYo), *Saddler Street*, Durham, and *Sadler Gate*, Derby, *Saltergate*, Chesterfield and Lincoln (making salt), also *Salter Row*, Pontefract, *Skinner Lane*, Bradford (WYo) and Leeds (WYo) (dealing in animal skins), *Smith Street*, Exeter (Dev) and Warwick (working in metal), *Spicer Street*, St Albans (Her) (meaning more generally 'grocer' rather than 'dealer in spices'), *Tanner Street*, Barking, east London, *Weaver Row*, Ipswich (Suf), *Wheeler Gate*, Nottingham (ie wheelwright) and *Woolmonger Street*, Northampton.

Most such trades are medieval, and the names date back accordingly. *Fishergate*, Norwich, is first recorded (as *Fischergate*) in 1285, for example, while *Pottergate*, Lincoln, is known from the 12th century, spelt exactly as now.

In some instances, however, the name may preserve an occupation in a form or spelling that can obscure its original sense. *Acreman Street*, Sherborne (Dor), is thus named for a farmer or ploughman (literally an 'acre man'), while *Barker Gate*, Nottingham, and *Barkers Pool*, Sheffield (SYo), refer to a 'barker', ie a tanner. *Baxter Gate* in Hedon (Hum) and Doncaster (SYo) refer to a baker, from Old English *bæcestre*, originally the feminine form of the word. *Bearward Street*, Northampton is named for a 'bear ward', that is, a keeper of performing bears, a popular sight in medieval times. *Bellar Gate*, Nottingham, may refer to a 'bell ward', a 'keeper of the bell', in other words, a town crier. Bells certainly lie behind the name of *Billiter Street*, City of London, recorded as *Belyeterslane* in 1298 and referring to a bellfounder. *Birchin Lane*, also in the City, may have a name relating to a barber, from an unrecorded Old English word *beardceorfere*, literally 'beard carver'. *Blake Street*, York, probably derives its name from Old Norse *bleikja*, 'bleach', and so refers to the bleaching carried on here. *Bleachfield Street*, Alcester (War), more certainly takes its name from that of the field where linen was regularly bleached. *Blossomgate*, Ripon, NYo, and *Blossom Street*, York, appear to refer to florists or flowers. Early records of the names show, however, that the meaning is really 'ploughswain street', from Old Norse *plógr*, 'plough' and *sveinn*, 'young man', 'swain' (as in modern *boatswain*). The reference is perhaps not so much to a ploughman as to a maker or repairer of ploughs.

Blowhorn Street, Marlborough (Wil), first recorded in 1526, may take its name from a trumpeter who lived here. *Bridlesmith Gate*, Nottingham, refers to a maker of bridles, in a word that is not regularly recorded. (The unfamiliarity of the term perhaps led to the misprinting of the name as *Bridgesmith Gate* in the Yellow Pages *Street Maps of British Towns*; see Bibliography, p. 215.) *Cannon Street*, in the City of London, refers not to canons of the church or even gunfounders but candlemakers. The name is a corruption and shortening of *Candelewrichstrete* (1180), literally 'Candle-wright Street'. The expected development *Candlewick Street* thus did not follow. (Birmingham's *Cannon Street* may be an adoption of the London name.)

Catherine Street, Salisbury (Wil) has already been mentioned (in Chapter 6, p. 82). The name is a corrupt form of *Carter Street.* The altered form developed through *carterne*, the genitive plural of Middle English *cartere. Coppergate*, York, is another misleading name in modern terms. It derives from Old Scandinavian *koppari*, 'joiner', 'turner' (related to modern English *cooper*). The name was recorded in 1120 exactly as now. *Crocker Street*, Newport (IoW) refers to a 'crocker' or potter (who makes crockery), while *Cumbergate*, Peterborough (Cam), takes its name from the woolcombers who lived or worked there next to the skin market. *Felter Lane*, York, is named for the makers of felt here at one time. It is now *Fetter Lane*, doubtless under the influence of the London street (see Chapter 3, p. 48).

TRADING AND OCCUPATIONAL NAMES

At first sight the name of *Fletcher Gate*, Nottingham, might be taken to derive from the *fletchers* or arrow-makers who once lived here. But in 1335 the name is recorded as *Flesshewergate*, showing that it was originally the street of the 'flesh-hewers', that is, butchers. In any case *fletcher* is a word of Old French origin, first cited by the *Oxford English Dictionary* only in 1400. Salisbury (Wil) has *Gigant Street*, perhaps referring to a fiddler who lived here, from Middle English *gigour*, related to modern *jig*. The name is pronounced as if 'Jigant'. *Gildencroft*, Norwich, takes its name from Old English *gildenacroft*, 'croft of the guild brethren', referring to some kind of social or religious fraternity. The Anglo-Saxons do not appear to have had trade or craft guilds, although religious fraternities of guild brethren are known to have existed from the latter half of the 11th century. *Gluman Gate*, Chesterfield (Der), is nothing to do with glue but is a form of Old English *glēomann*, from *glīw*, 'play' 'sport', 'music', later giving *glee* as a form of unaccompanied part-song as well as a general word for 'pleasure'. *Iron Gate*, Derby, relates to the smiths or workers in iron here in medieval times, but *Knifesmithgate*, Chesterfield, may derive from a personal name rather than being a street where knifesmiths worked. *Lister Gate*, Nottingham, was formerly the street of the *litsters* or dyers, from a Middle English word to which modern *litmus* is related. *Longsmith Street*, Gloucester, is so named from its smiths or metalworkers, with the prefix distinguishing this street from the city's *Broadsmith Street*, no longer in existence. *Pagisters Road*, Abingdon (Oxf), is said to be so named for the *pargeters* or plasterers who formerly lived here.

The unusual name of *Petty Cury*, Cambridge, often causes speculation. It means 'little kitchen', from Old French *petit*, 'little' and *curie*, 'kitchen'. The street where the cooks lived and worked was probably so named to distinguish it from some other, larger *Cury*. (The basic word existed in English in the sense 'cookery', 'kitchen' down to the 16th century. *Petty Cury* is itself first recorded in 1330, as *parva Cokeria*.) Nottingham's *Pilcher Gate* takes its name from its pilchers, who made *pilches* (garments made of skin and fur). The name of *Rack Street*, Exeter (Dev), refers to the racks on which cloth was stretched (compare *Tentergate*, below), while *Skeldergate*, York, preserves the memory of its former shield-makers, from Old Scandinavian *skjaldari*, literally 'shielder'. *Souters Lane*, Chester, refers to the shoemakers here, from Old English *sūtere* or Old Norse *sútari*, literally 'sewer' (related to modern *suture* and eventually *sew* itself). *Souttergate*, Hedon (Hum), is of the same origin. Shoemakers or cobblers are indirectly alluded to in the name of *St Crispin's Road*, Norwich. The street is a modern one, but St Crispin is the patron saint of shoemakers and the name is appropriate generally for Norwich as a city noted for its footwear industry.

Spiceal Street, Birmingham, is named for its *spicers*, this word meaning 'grocer' before it gained its more specialised sense. *Spurriergate*, York, is so called for its *spurriers* or spur-makers. *Stramongate*, Kendal (Cum), is thought to preserve a reference to a *strawman*, that is, a dealer in straw. The word is probably of Middle English origin and is not recorded generally for this particular occupation. *Tentergate*, Knaresborough (NYo), takes its name from the *tenters* used for stretching cloth after milling (compare *Rack Street*, above). (The same word lies behind the expression *on tenterhooks*, the latter literally being the hooks on which the cloth was fastened for stretching on a tenter.) The name of London's famous *Threadneedle Street*, the home of the Bank of England in the City, is said to be a corruption of *three needles*, the sign of the needle-makers formerly here. Some, however, see the origin in an old children's game called *thread the needle*, perhaps played here at one time.

Tilehouse Street, Hitchin (Her), seems to denote a street where tiles were made, or where tilers lived. *Tucker Street*, Bristol, was known for its *tuckers* or cloth finishers. An identical occupation is preserved in the names of the same city's *Walker Street*, found elsewhere as *Walkergate* in Beverley (Hum) and Pontefract (WYo) and *Walker Lane*, Derby. Both *tucker* and *walker* are terms for a fuller, who beat or pressed woollen cloth in order to shrink and thicken it. He originally did this by *walking* or stamping on it. *Waterbeer Street*, Exeter (Dev), seems to take its name from local water bearers or carriers, while *Wolsdon Street*, Plymouth (Dev), is recorded as *Woolster Street* in 1637, so apparently relates to the *woolsters* or dealers in wool here.

Occupations that are more literary than manufactory are alluded to in the names of *Doomgate*, Appleby (Cum), where trials were held (from Old English *dōm*, 'judgement'), *Printing House Square*, in the City of London, occupied by the King's Printers in the 17th century and later the site of the first publication of *The Times*, and *Scholars Lane*, Stratford-on-Avon (War), doubtless referring to the scholars of the Guild School here. This last was formerly known as *Tinker Lane* (*Tynkereslone* in 1331) for its *tinkers*, or menders of pots and pans, so has gone up in the world.

Chancery Lane, in the City of London, is named more for an office than an occupation. It was originally *New Street* (recorded as *nouus vicus* in 1185), then *Chancellor's Lane* (*Chauncelereslane* in 1338). The present name is first recorded in 1454 (as *Chauncerylane*). The name has its origin in the office of Master of the Rolls of Chancery set up here in 1290 in a former religious house. (The office itself originated as that of the guardian of all the charters, deeds and the like that had been entered on parchment rolls in Chancery, that is, in the court of the Lord Chancellor.) The name *Chancellor's Lane* seems to suggest that the Lord Chancellor actually lived in the street. It does not necessarily follow that *Chancery Lane* developed as a name from *Chancellor's Lane*. It could have evolved independently, with the name of the office replacing the title of the officer.

It will be noticed that several of the occupations here have given modern surnames, such as Baker, Barker, Baxter, Brewer, Carter, Fisher, Potter, Skinner, Smith, Spicer, Tucker, Walker and Wheeler. In some instances the street may thus have been named after a person who lived there or who owned property there. In the majority of cases, however, the streets have been named directly for the various occupations of the craftsmen and traders who had houses or premises there.

Some streets are named not for the different occupations of their inhabitants but for their nationalities.

Bartholomew Street, Exeter (Dev), was originally *Breton Street* (*Britaynne* in the mid-13th century), so called from Bretons who settled here. The present name may have been suggested by the old. An identical name is that of *Breton Side*, Plymouth (Dev). It may be significant that Brittany is due south of both towns across the English Channel. *Little Britain* in the City of London, however, may be so called from a surname such as *Le Breton* rather than for the nationality. *Danesgate*, Lincoln, is so named for the Danes and for the Danelaw established by them in northern and eastern England, including the region in which Lincoln lies. The name is first recorded, as now, in 1180. It then fell out of use, but was revived in the 19th century by the city's Lighting and Paving Commissioners. *English Street*, Carlisle (Cum), led to the *English Gate*, as the south gate of the city. It is said to be so named as it was one of the parts of the city in which 'dwelt the chief and best citizens, natural Englishmen' (J. Denton, *An Accompt of the most considerable Estates and Families in the County of Cumberland*, 1887). Carlisle's present *Castle Street* was similarly known originally as *via Francorum*, 'street of the French'.

Flemingate, Beverley (Hum), was named for the Flemish merchants who traded in the town, while *Frenchgate*, Richmond (NYo), refers to its former French inhabitants. *French Gate*, Doncaster (SYo), does likewise, with the name first recorded in the 12th century. (Compare *Scot Lane*, below, in the same city.) *French Row*, St Albans (Her), is another name deriving from this nationality. The street must have been occupied by French merchants and shopkeepers. *Gipsy Hill* can be included here, although not relating to a specific nationality. The street so named in Lambeth, south London, recalls the gipsies who lived in this part of former Surrey from the 17th century. *Greek Street*, west of Charing Cross Road, London, is named for the Greek Orthodox church that was built in the late 17th century for Greeks who had sought refuge here from Turkish rule. *Landress Lane*, Beverley (Hum), is more a local name, but can also be admitted here. It was recorded as *the Londiners street* in 1660, and apparently refers to the merchants from London who held a fair in Beverley in the 17th century.

London itself is famous for *Lombard Street*, in the City. It represents the site of the wool market held here in the 13th century by Italian merchants from Lombardy. They formed the nucleus of the merchant bankers with whom the street is still associated today. The *Rue des Lombards* in Paris is a street name of identical origin. *Petty France*, Westminster, west London, is another well-known name. It means 'little France', referring to the French merchants who lived and traded here in medieval times. In the late 18th century the street was renamed *York Street*. In 1920, however, the old name was restored thanks to the initiative of Sir Laurence Gomme (1853-1916), a founder of the Folklore Society and Clerk to the London County Council.

Sardinia Street, Holborn, north London, is so called from the chapel of the Sardinian Embassy that stood here from the mid-17th century to the early 20th. *Scot Lane*, Doncaster (SYo), is said to refer to the Scotsmen who once lived here. As such, the name complements that of *French Gate* (above). *Sicilian Avenue*, off Southampton Row, central London, is not named for its inhabitants but for the Sicilian marble with which it was paved on its construction in 1905. Petersfield (Ham) has a street named *The Spain*, popularly said to commemorate Spanish merchants who came here to buy wool or cloth. Supporting evidence for this origin is lacking, however, and the name may be a corruption of some kind. *Welsh Row*, Nantwich (Che), on the other hand, is better documented. It preserves the memory of the Welsh merchants who came to Nantwich for the salt trade. The original name of the street was *Frog Row*, referring to the prevalence of these creatures as a result of the location of the street on the west bank of the River Weaver.

A few names derived from religious groups belong here rather than in Chapter 6. *Huguenot Place*, Wandsworth, southwest London, is so named for the Protestant Huguenot refugees from religious persecution in France who settled in Wandsworth in the late 17th century. The street adjoins the Huguenot cemetery here. On similar lines *Moravian Place*, Chelsea, southwest London, refers to a Moravian religious community who had their home here a century later. Here, too, may be included *Maids Causeway*, Cambridge, although the religious connection in this case is more tenuous. The street (mentioned in Chapter 3) is said to be so named from the almshouses founded here in 1647 for 'two widows and four maids', that is, for two poor women who had been married and four who had not.

The most common name of this type, however, refers to the Jewish community, and is found for *Jewbury*, York, *Jewry Street* and *Old Jewry*, City of London and *Jury Street*, Warwick. A name of identical reference but more unusual origin is *Jubbergate*, York. The street, one of the first to be built outside the city walls, was originally known as *Bretgate* (*Brettegata* in 1145), 'street of the Britons'. This referred to the Cumbrian Britons who had settled in northern England with the Irish Vikings from the 8th century. The present name is first recorded in 1356 as *Jubretegate*, representing the same name but with *Jew* prefixed. The addition relates to the Jews who had come to live in this street after Edward I had driven them from the city.

Finally, mention should be made of a name that is not occupational or even strictly speaking religious. *Eden Street*, Kingston upon Thames, southwest London, almost certainly derived its name as a corruption of *Heathen Street*, as which it is first recorded in 1353. The precise application of the name remains obscure, however.

8. NAMES OF BUILDINGS AND STRUCTURES

The preceding chapters have instanced many street names derived from buildings, such as the churches and religious foundations included in Chapter 6 and the trading and manufacturing places mentioned in Chapter 7.

As a basically urban and 'manmade' feature, it is hardly surprising that a street should have a name of this kind, if only because one primary function of a street is to provide access to a building.

In a few cases, the building that gave its name to a street is referred to only in general terms. This is so for *Bold Lane*, Derby, where the name probably represents Old English *bold*, 'building'. Specifically new buildings are referred to by the names of *Newbegin*, Beverley (Hum) and *Newbiggin* Street, York, where the second element of the name represents Middle English *bigging*, 'building', itself ultimately related to *be* (but not to *big*).

When it comes to a particular type of building, however, the reference is frequently to a historic defensive structure of some kind, ranging from a city wall to a fort. *Aldwark*, York, thus has a name that means literally 'old work', that is, 'old fortification'. The street runs parallel to the present city wall and doubtless marks the course of the earlier Roman wall. City walls or defences have given names elsewhere, such as the self-descriptive *City Walls Road*, Chester, *Longwall Street*, Oxford, similarly running parallel with the city wall, and *Rampart Street*, Salisbury (Wil), from the old city ramparts, portions of which survived to the mid-18th century. *Warser Street*, Nottingham, is believed to derive its name from a combination of Old English *weall*, 'wall', and *setu*, 'buildings'.

London's *Barbican*, now thought of chiefly as a cultural centre or residential complex, gave the name of *Barbecanstret*, recorded in the mid-14th century, and served subsequently as an elliptical street name (without *Street*). The reference is to the barbican or walled outwork that protected the entrance to the City of London in medieval times. The ultimate origin of the word itself is unknown, but it is not unique to London and is also found, for example, in *Barbican Road*, Barnstaple (Dev), east of the town centre. There was likewise a street named *Barbican* by the city walls of Exeter (Dev), recorded as *the Barbycan* in 1562 but no longer current.

Old English *burh*, which gave modern *borough*, meant 'fortified place' or, later, 'manor house', and is also found behind some street names. Examples include *Bury Road*, Hemel Hempstead (Her), *Buryfields*, Guildford (Sur) and *Sidbury*, Worcester, this last meaning 'southern fortification'. *The Borough*, Petersfield (Ham) and *Borough High Street*, Southwark, south London, on the other hand, are later names, the latter referring to the Borough of Southwark. In Scarborough (NYo) *Newborough* probably refers to the (now ruined) Norman castle towards which it leads. Together with *Westborough* and *Eastborough* it forms the town's single central thoroughfare. *The Forbury*, Reading (Ber), appears to belong here, since Old English *fore-burg* meant 'outwork'. But archival evidence makes it clear that the name actually referred to the outer court of Reading Abbey, so that the sense is more exactly 'forecourt'.

Both *Mount Street*, Westminster, west London, and *Mount Terrace*, Stepney, east London, have names relating to Parliamentary forts built in the Civil War (1642-8). *Vastern Road*, Reading (Ber), refers to an older defensive structure, dating from medieval times. The name derives from Old English *fæst-ærn*, 'fortified house', perhaps one set up at the northern end of the rampart built by the Danes in the 9th century from the Kennet to the Thames.

Bailey, as a term for the outermost wall or court of a castle, is found in several street names. Examples include *The Baile*, York, *The Bailey*, Skipton (NYo) and Worcester, *North Bailey* and *South Bailey*, Durham, *North Baileygate*, Pontefract (WYo) and, most familiarly, *Old Bailey*, in the City of London. The last-named arose as a street running beside an outwork on the

City wall. It was known simply as *The Bailey* to the 15th century.

Tower Hill, Stepney, east London, is so named from the famous Tower of London, whose original White Tower, built in the 11th century, gave the name of the fortress as a whole. Similar names elsewhere include *Tower Ramparts*, Ipswich (Suf) (earlier *Tower Ditches*, then *Tower Terrace*), referring to the northern rampart area of the town (which has no castle), and *Tower Hill* and *Tower Street*, Bristol, for the tower of Bristol Castle. *Lucy Tower Street*, Lincoln, takes its name from the small tower at the south end of the western city wall, itself named for *Lucy Tower* as the keep of Lincoln castle, from Lucy 'the countess', who lived in the 12th century and who is claimed by some to have been the daughter of Thorold, sheriff of Lincolnshire. *Citadel Way* and *Tower Street*, Hull (Hum), are two modern streets east of the estuary of the River Hull. They take their names from the triangular area of land here known as the Citadel where the army garrison was formerly located (compare *Garrison Road*, below, p. 119.) A related name is that of *Sentry Road*, Swanage (Dor), where a military lookout post was situated.

The Brittox, Devizes (Wil), is an unusual street name referring to a place that was fortified with stockades, in this case perhaps the route to the old castle gates. It was recorded in about 1300 as *la Britasche*, and has its origin in an Old French word related to modern English *brattice*, now used as a term for a wooden tower or parapet in medieval fortifications. Almost as unusual is the (much more modern) name of *Warrenne Keep*, Stamford (Lin), a private road leading to a modern apartment complex. It is located on the site of the circular keep of the former Norman castle here, with the first word of the name referring to William de Warenne (*sic*), Earl of Surrey, who held the manor of Stamford in the early 13th century.

Some streets have come to be named for a group of buildings, such as a row of houses. *Barum Top*, Halifax (WYo), takes the first word of its name from Old English *būrum*, 'at the cottages', and *The Headrow*, Leeds (WYo), refers to a row of houses at the head or uppermost part of the town. *Raw Gap*, Knaresborough (NYo), is a street that emerged in the gap of a row of houses, while *The Rows*, Chester, in the centre of the city, is the name of the rows of buildings and shops that have become attractive to tourists for the public footways inside their first-floor fronts. The earliest record of this latter name is as *communis via vocata Lofts* (1492), 'the public way called The Lofts'. *Little London*, Chichester (WSx), is probably a name that belongs here, referring (perhaps semi-humorously) to the many houses built in the 19th century between East Street and Priory Park. *Stepcote Hill*, Exeter (Dev), is a similar 'group' name, relating to the cottages or houses that line the series of steps here.

Streets named for medieval manors are among the oldest on record. *Allfarthing Lane*, Wandsworth, southwest London, is one such, with the manorial name meaning 'half a fourth part', probably referring to an eighth part of an earlier grant of land. *Basing Court*, Camberwell, southeast London, commemorates a manor held by a family who came from Basingstoke (Ham). *Boroughbury*, Peterborough (Cam), took its name from the old manor of *Burghbury*, that is, 'Peterborough Bury'. This historic name has now been subsumed into that of *Lincoln Road*, the former main road to Lincoln before the construction of *Bourges Boulevard* in the 1970s. Less obviously, *Bridge Court*, Battersea, south London, is named for a lost manor that itself took its name from a bridge over the Falcon Brook here, now covered over. *Caldecott Close*, Abingdon (Oxf), is named for a manor in Sutton Wick whose own name means 'cold cottage'. *Dowlas Street*, southeast London, is named for another Camberwell manor, itself named for John de Uvedale, who held it in the early 14th century.

Crossbrook Street, the central street of Cheshunt (Her), is named for the former manor here, whose own name means 'cress brook'. *Ebury Street*, in Westminster, west London, has a manorial name meaning literally 'island bury', with reference to the former marshy land here. The manor represented the western part of the present City of Westminster, and included Hyde Park. *Kingsbury Square*, Wilton (Wil), and *Kingsbury Street*, Calne (Wil), are both named for former royal manors. *Levehurst Way*, Lambeth, south London, commemorates a manor whose own name meant 'Lēofa's wooded hill'. *Lisson Grove*, Marylebone, northwest London, bears a manorial name meaning 'Lille's farm'. *Milkwood Road*, Dulwich, southeast London, probably preserves the old manorial name *Milkwell*, recorded from the late 13th century and meaning 'milky well', referring to the colour of spring water here. *St Ives Road* and *Shoppenhangers Road*, both in Maidenhead (Ber), are likewise named for old manors. The former, representing a personal name, had *St* prefixed in the 19th century on a popular analogy with St Ives. The latter means 'Sceobba's wooded slope'. Oxford's *Walton Street* originally led to the manor of this name, probably meaning 'homestead by the wall'. *Westbury*, Sherborne (Dor), is a street with a name meaning originally 'western manor'. The place-name *Eastbury*, as that of the 'eastern manor', is now lost here, although it has been preserved in that of the *Eastbury* Hotel, Long Street.

A number of streets take their names from a relatively recent house or mansion. The house itself may have a contemporary name or have adopted it from a manorial or other medieval location. *Ashburnham Road*, Chelsea, west London, is named for Ashburnham House, itself named for the second Earl of Ashburnham, who lived here in the late 18th century. The house survived until about 1880. *Bedfordbury*, Westminster, west London, takes its name from a house built here in 1601 by the Earls of Bedford. *Bradshaw Street*, Manchester, commemorates the mansion known as Bradshaw Hall, for many years the home of the family of the same name. The house gradually fell into decay and was pulled down piecemeal in the early 20th century. *Burleigh Road*, Loughborough (Lei), bears the name of the 17th-century Burleigh Hall, demolished in 1961. *Cambray Place*, Cheltenham (Glo), preserves the name of a house described in

1805 as 'an elegant mansion newly built on Cambray meadow'. (The field name represents that of a local family called *Cambray*, their own name ultimately deriving from *Cambrai* in France.)

Campden Hill, Kensington, south London, was so named for Campden House, built here in the early 17th century by Sir Baptist Hicks, later Lord Campden. The house was destroyed by fire in 1862. London's well-known *Carlton House Terrace*, in the City of Westminster, is clearly derived from a house name, that of Carlton House, built here in 1709 for Henry Boyle, Lord Carleton. The mansion was rebuilt for the Prince of Wales in 1772, but was demolished in about 1820, when it was replaced by the present terrace. (See also Appendix, p. 195.) Equally famous, though for reasons of fashion rather than fortune, is *Carnaby Street*, also in Westminster. It takes its name from *Karnaby House*, owned here by one Richard Tyler, a property developer, in the 17th century. The origin of its name is unknown, though there may be some link with the village of *Carnaby* near Bridlington (Hum), whose own name was frequently spelt with *K-* in medieval times.

Colliton Street, Dorchester (Dor), leads to Colliton House, itself bearing the name of its location, meaning 'Coll's homestead'. *Cremorne Road*, Chelsea, west London, is named for the pleasure gardens that were here on the site of the former Cremorne House, bought by Lord Dartrey, later Viscount Cremorne, in 1778. *Cromwell Road*, Kensington, south London, has the name of Cromwell House, demolished in the mid-19th century. The house, originally known as Hale House, was so called from a tradition that Henry Cromwell, son of Oliver Cromwell, lived here at one time. *Figtree Lane* and *Mulberry Street*, both in Sheffield (SYo), are named for houses that were themselves so called from their trees. *Folly Lane*, St Albans (Her), bears the name of *Folly House*, also known as Bleak House, itself perhaps so nicknamed for its extravagant size or style. *Garendon Street*, Loughborough (Lei) is named for the now demolished Garendon Hall, home of the De Lisle family, while *Hawthorn Street*, *Hawthorn Grove*, *Hawthorn Lane* and

Hawthorn Walk, Wilmslow (Che), all commemorate Hawthorn Hall, itself named for the hawthorns that grew here.

Rather unexpectedly, *Helen Road*, Weymouth (Dor), is so called after a former poor tenement here nicknamed *Hellhouse*. On the other hand, *Jamaica Road*, Bermondsey, south London, accurately preserves the name of a Jacobean house here which itself was so called for the island that became a British colony in 1655. *Little Britain*, Dorchester (Dor), is said to have adopted the name of a former mansion known as *Britain* by the River Frome. The house was probably so called after a field here, itself named for a family who originally came from *Brittany*. *Loughborough Road*, Lambeth, south London, has preserved the name of Loughborough House, the home of Henry Hastings, created Lord Loughborough in 1643. The gardens that gave the name of *Norham Gardens*, Oxford, were at one time part of the estate of Norham Manor here.

It will be noticed that many mansions are named for their owners. One such gave the name of *Orleans Road*, Twickenham, southwest London. Orleans House here was so called after Louis-Philippe, Duke of Orleans, who lived here in exile from 1815 to 1870. Most of the house was demolished in the 1920s, although the Octagon survives. *Pavilion Road*, Chelsea, west London, commemorates a house called *The Pavilion* which stood here for just over a century from 1777.

London's famous *Piccadilly* derives its name from a 17th-century house here nicknamed *Piccadilly Hall*, so called from the fact that its owner, a tailor named Robert Baker, had made his fortune through the manufacture of *piccadills*, a highly fashionable type of wide high collar. By a curious coincidence *Piccadilly Circus*, at its eastern end, stands on the site of a house whose garden formerly lay by a field known as *The Round Ringill*. The latter word is a form of *ringell* (modern *roundel*), meaning 'small ring', so that the present Circus happens to reduplicate the former sense almost exactly.

Ranelagh Grove, Chelsea, west London, takes its name from the pleasure gardens that were laid out on the estate of Richard

Jones, first Earl of Ranelagh, who built his house here in 1690. The site is now that occupied by the Royal Chelsea Hospital. The same pleasure gardens also gave the name of *Ranelagh Avenue* and *Ranelagh Gardens*, Barnes, west London, as these streets took their name from a club at Barn Elms which had itself been named for the gardens. The Earl took his Irish title from Ranelagh, now a district of Dublin.

Silvertrees Drive, Maidenhead (Ber), was so named for a house that was demolished when the street was built. The town council had wanted to name the new road *Tuggins Close*, from a long-established local name, but abandoned this following complaints from would-be residents. *Southcote Road*, Reading (Ber), is so named for Southcote House, which was itself so called after the former 'southern cottages' here. *Stafferton Way*, also in Maidenhead, was laid out in the 1930s in the grounds of Stafferton Lodge, itself having a medieval manorial name. *Thurland Street*, Nottingham, is similarly named for a house formerly here named either Thurland House or Thurland Hall. The mansion was itself named from the family of one Thomas Thurland, to whom there is a 15th-century memorial in nearby St Mary's church.

Dollar Street, Cirencester (Glo), is so named not for a private house but for a 'dole hall', where charitable gifts were made to the poor. The present form of the name evolved by association with the American currency. Another public building gave the name of *Hall Gate*, Doncaster (SYo), so that the meaning is simply 'hall street'. The name is first recorded (as *Hawegate*) in 1409. *Markhouse Road*, Walthamstow, northeast London, is named for a former house called *Markhouse*. It took its name from the medieval manor of Mark, itself so called because it straddled the boundary (Old English *mearc*) between Leyton and Walthamstow. *Stonehouse Green*, Congleton (Che), more obviously takes its name from a stone-built house here at one time. It was doubtless actually known as 'the Stonehouse'.

Not all streets with *Manor* in their name are so called with reference to a medieval manor. *Manor Road*, Swanage (Dor), for example, has a recent name, from a hotel that was originally called 'Manor House' but that in 1835 was renamed 'The Victoria Hotel' after the 16-year-old Princess Victoria had stayed here on a visit to Swanage. *Manor Road*, Oxford, too, is named for Holywell Manor, which although itself of medieval origin was rebuilt in 1516.

Some streets are named specifically for a *palace*, or at any rate for buildings called 'Palace'. An obvious example is that of London's *Buckingham Palace Road*, leading to the famous royal residence. *Gaunt Street*, Lincoln, takes its name from the so-called John of Gaunt's Palace here, and *Palace Street*, Norwich, is more obviously named for the Bishop's Palace. Less transparent is the name of *Surrey Street*, Norwich, so called after Surrey House, a palace built in the mid-16th century by the poet Henry Howard, Earl of Surrey. Nothing is left of the palace now, and the Norwich Union Building stands on the site.

One of London's best known streets, *Whitehall*, is also named for a former palace. This was originally York Place, the London residence of the Archbishops of York, which had been enlarged and improved by Cardinal Wolsey when he himself became Archbishop in 1514. After his downfall in 1529 the palace was appropriated by Henry VIII and renamed *White Hall*, either from the colour of its stone or from the traditional name given to any kind of festive hall. (There is another, also known as the 'Great Chamber', at Kenilworth Castle (War).) All but the Banqueting Hall of Whitehall Palace was destroyed by fire in 1698, though its existence is preserved in the name of the street, now more readily associated with government offices.

Sometimes a mill has given the name of a street. One such is *Coppermill Lane*, Walthamstow, northeast London. *Great Windmill Street*, in the City of Westminster, London, is named for a windmill recorded as being here in the 16th century. The street is *Great* by comparison with the former *Little Windmill Street*, now *Lexington Street. Iron Mill Road*, Wandsworth, southwest London, is so named from a foundry by the River Wandle here at one time. London's *Millbank*, extending from

Vauxhall Bridge to Victoria Gardens, is named for a watermill that belonged to the Abbey of St Peter, Westminster, and that is recorded as being here to at least the mid-17th century. *Mill Street*, Macclesfield (Che), on the other hand, is named for the first silk mill to be set up in the town, in 1756. *New Mills Yard*, Norwich, is despite its name not a yard at all but a street, running from Westwick Street over the River Wensum to Oak Street. Its name refers to the corn mills which were 'new' as long ago as 1430.

Ray Mill Road, Maidenhead (Ber), is named for a mill formerly here that stood on land known as *Ray*, apparently from Old English *æt thǣre ēge*, 'at the island', or else directly from a stream here called the *Ray* that itself gained its name from this source. The same name has given those of many streets in this district of Maidenhead, including not only *Ray Mill Road* (now divided into *Ray Mill Road East* and *Ray Mill Road West*) but also *Ray Drive*, *Ray Lea Close*, *Ray Lea Road*, *Ray Mead Road*, *Ray Park Avenue* and *Ray Park Road*. *Milford Lane*, Westminster, west London, is said to have a name that refers to a watermill formerly here that stood on a fast-running stream flowing down to the Thames. *Windmill Hill*, Hampstead, northwest London, takes its name from the medieval flour mill that belonged to the manor of Hampstead, while *Windmill Road*, Wandsworth, south-west London, has a name referring to a wind-operated water-pump that was set up here in the early 19th century.

Streets bearing mill names are complementary to the trading and occupational names mentioned in Chapter 7. Other names derived from buildings used for manufactory purposes may also be included here.

Mint Street, Southwark, south London, takes its name from the royal mint that operated here in Suffolk Place in the 16th century, while *Mint Street* in Lincoln is said to preserve the name of Mint Hall, a building connected with the former Lincoln mint. The site of the building is now that on which (appropriately) the National Westminster Bank stands. *Royal*

Mint Street, Stepney, east London, commemorates the Royal Mint itself, originally operating in the Tower of London but set up at the end of this street in 1811. In 1968 the Royal Mint was moved out of London altogether to a larger site in Llantrisant, south Wales.

Gas Street, Lincoln, is one of a number of identically named streets still found in many towns in the country. Here, as elsewhere, the name refers to the nearby gas works. Most streets of this name date only from the 19th century. *Waterhouse Lane*, Hull (Hum), is so called from the town's waterworks, which were built at the northern end of the street in the 17th century.

Union Street, Hemel Hempstead (Her), is named for the union workhouse that was built here in 1837 and that has a modern successor in St Paul's Hospital. Workhouses of this type were so called because they were run by a *union* of local parishes formed under a single board of guardians to administer the poor laws. *Corporation Row*, Islington, north London, has a similar name, derived from the New Corporation Workhouse that was opened here in about 1665. *Malthouse Road*, Norwich, is a reminder that the city had several malthouses in the 19th century, with two on *Malthouse Lane*, as the street was then called.

Business premises of one sort or another also give street names. An example is the self-explanatory *Dock Office Row*, Hull (Hum). *Exchange Court*, in the same city, takes its name from the Exchange and News Room opened nearby in 1794. *Exchange Street*, Manchester, on the other hand, commemorates the commercial exchange built in 1729 but superseded by the famous cotton trading centre, the Manchester Royal Exchange, in 1809. *Exchange Street*, Norwich, by contrast, is named for the corn exchange that opened in 1828 in Bedford Street.

Great Clarendon Street, Oxford, is named for the Clarendon Press, the printing house of the Oxford University Press. It was originally *Clarendon Street*, but was prefixed *Great* because it is longer and wider than nearby *Little Clarendon Street*, formerly *Blackboy Lane*. The press is now in Walton Street, but until 1832 was here in the Clarendon Building, erected in the early 18th century on proceeds from the publication of Lord Clarendon's *History of the Rebellion* (1704).

A more rural workplace and 'factory' is a *farm*, as reflected in the names of a number of streets, although not always obviously. London's *Aldwych*, in the City of Westminster, has a name meaning 'old farmstead', from Old English *ald*, 'old' and *wīc*, 'farm', often specifically implying a dairy farm. The street was so named by the London County Council in 1903. Today it is often called '*the* Aldwych', as if the original meaning were sensed.

Catlins Lane, Ruislip, west London, is so called from a farm originally known as St Catherine's End Farm, and the present name is a corruption of this. *Grange Road*, Peterborough (Cam), is a modern road laid out in the 1920s over the former Westwood Grange farm. The same farm gave the name of *Westwood Park Road* here. *Hay Hill*, Westminster, west London, takes its name from a farm formerly here. The farm's own name does not derive from *hay*, however, but from Middle English *eye*, representing Old English *ēg*, 'island', implying land that is well watered, perhaps in this case by the Tyburn which at one time flowed here. The very small *Longhedge Street*, off Rowditch Lane in Battersea, south London, preserves the name of the former Long Hedge Farm, recorded here in the mid-17th century. The farm was enclosed by a 'long hedge' that ran along what is now Battersea Park Road. The farmhouse survived until 1965. *Susans Road*, Eastbourne (ESx), does not bear a personal name but that of Susans Farm, itself preserving the Medieval Latin phrase *terra susana*, 'worn-out land'. In Oxford, *Oxpens Road* is a fitting name in that city for a street named after a farm where oxen were kept.

Some streets are named not for an entire farm but for farm buildings or premises. *Barn Hill*, Stamford (Lin), is so called from a large barn formerly here. *Feasegate*, York, is a name that means 'cowhouse lane', from a contracted form of Old Norse *fé-hús*, 'cow house'. *Laith Gate*, Doncaster (SYo), has a name meaning 'barn street', from Old Norse *hlatha*, 'barn', to which is related the north of England dialect word *lathe* in the same sense. (The word is ultimately akin to modern English *laden* and *load*.) The Doncaster street is now divided into *East Laith Gate* and *West Laith Gate*. In Gloucester, *The Oxbode* is the name of a street that formerly led (as *Oxbode Lane*) to an ox house.

Streets named for hospitals can derive their names from a medieval house of religious origin or from a more recent establishment. *Hospital Street*, Nantwich (Che), has already been instanced (in Chapter 6) as an example of the former type. *Hospital Road*, Sherborne (Dor), on the other hand, is so named for the Yeatman General Hospital (formerly the Yeatman Memorial Hospital), opened by public subscription in 1865. It was originally known as *Back Lane*, now the name of a shorter adjacent street.

A number of towns have a street named for a local school. *Burlington Road*, Fulham, west London, is named for a school called Burlington House that was opened here in the late 18th or early 19th century by a Mr Roy, who came from Old Burlington Street, Piccadilly. *Challow Court*, Maidenhead (Ber), is named from a private school here at one time. *Greycoat Place*, Westminster, west London, bears the name of Greycoat Hospital, founded as a school for 40 poor boys and 40 poor girls in 1695 and still here today as a Church of England comprehensive school for girls with over 900 pupils. *Grammar School Road*, Hull (Hum), was so named in 1937 when a site was acquired for new premises for Hull Grammar School. However, the school moved only in 1953, and not to this road but to *Bishop Alcock Road*, named for John Alcock, Bishop of Ely, the school's chief benefactor in the 15th century. The name is thus a misnomer. It remains, however, despite proposals to change it to something more appropriate. *Schoolhouse Lane*, Stepney, east London, is truer to its origin, and has a name referring to the Coopers' Company's School, founded here in the first half of the 16th century. The school, today known as the Coopers' Company's and Coborn School, a boys' comprehensive with over 1000 pupils, is now located in Upminster, northeast London.

Some streets in Oxford and Cambridge are named for the college past which they run, such as *King's Parade*, Cambridge, already mentioned. In a few cases, however, the college no longer exists, or the street has been renamed. *New Inn Hall Street*, Oxford, thus preserves the name of New Inn Hall, a college absorbed by Balliol College in 1887. St Peter's College now stands on its former site. Also in Oxford is *St Bernard's Road*. This was earlier *St John's Road*, after St John's College here. In 1961, however, the name was changed in order to avoid confusion with *St John Street*, further to the south, and also apparently named for the college. The new name is that of the college from which St John's developed. In Cambridge, *Sidney Street* is named for Sidney Sussex College. It was originally *Conduit Street*, a name first recorded (as *via del cundyt*) in 1293. *Senate House Passage*, Cambridge, runs by the Senate House, the chief administrative building of the university, where important public functions are held. It corresponds in this respect to the Sheldonian Theatre at Oxford, a building whose name has not been transferred to any street. Its founder, however, is commemorated by the modern *Sheldon Way*, in Cowley, southeast Oxford.

Logic Lane, Oxford, is named for a former school of logicians at its northern end. *Pusey Street*, in the same city, is named for nearby Pusey House, the Anglican theological library and chapel. Until 1925 the street was known as *Alfred Street*, although it is not certain which particular Alfred was commemorated. The name was doubtless changed to avoid confusion with the *Alfred Street* further to the south, joining the High Street and Blue Boar Lane. *Free School Lane*, Lincoln, is named for the school founded here in 1568 when the grammar school was moved to the Grey Friars. A 'free school' was one in which at least some of the pupils received their education without payment. *Garret Hostel Lane*, Cambridge, is a footpath over the River Cam leading to Trinity College. It derives its name not from a mythical St Gerard, as sometimes explained, but from a hostel with a high tower or window that served as a lookout post. The original hostel, recorded as existing here in the 15th century, was incorporated in the foundation of Trinity College. In London, *University Street* is the approach to University College London from Tottenham Court Road.

Streets named for military premises include *Barrack Street*, Warwick, named from the Militia Barracks here, *Barrack Street*, Norwich, so called from the former Horse or Cavalry Barracks, demolished in the 1960s, and *Garrison Road*, Hull (Hum), built in 1981 over the site of the former garrison in the region still known as the Citadel.

New Scotland Yard is today not a street name as such, but that of the headquarters of the Metropolitan Police Force in London. However, its history involves the names of streets, so deserves inclusion here. Part of the former Whitehall Palace (see *Whitehall*, above) had been occupied by visiting kings of Scotland and their ambassadors. When most of the palace was destroyed by fire in the late 17th century, however, the only memorial to the kings was the name of *Great Scotland Yard*, so called by contrast with the alleys known as *Middle Scotland Yard* and *Little Scotland Yard*, which were later combined into *Whitehall Place*. A house in the last-named was selected in 1829 for the headquarters of the newly formed Metropolitan Police Force. It backed onto the still existing *Great Scotland Yard* and this name, less the superfluous *Great*, was taken up as a nickname for the police offices as a whole. In 1890 the establishment moved from its cramped premises here to a new site on the recently completed Victoria Embankment, near Westminster Bridge, where it was officially named *New Scotland Yard*. This name was retained when the 'Met' moved yet again in 1967 to its present premises in Victoria Street.

On the opposite side of the coin, so to speak, are the streets named for prisons. *Bridewell Alley*, Norwich, takes its name from the *bridewell* or house of correction opened here for 'vagabonds and minor delinquents' in 1585. *Bridewell Street*,

Marlborough (Wil), is similarly named for the town prison, opened in 1787. *Bridewell* as a generic term for a prison derives its own name from the London penal institution which stood near *St Bride's* (ie St Bridget's) *Well* not far from Fleet Street, where it is itself commemorated in *Bridewell Place. Clink Street*, Southwark, south London, bears another generic term or nickname for a prison, in this case one catering for the troublesome frequenters of local brothels. *Goal Lane*, Halifax (WYo), preserves a reference to a former jail here. Modern *gaol* was frequently spelt *goal* in the 16th and 17th century. *Marshalsea Road*, Southwark, derives its name from the prison which formerly stood here from medieval times and which in its final years was notorious as a debtors' jail. The prison was so named as it was under the control of the Knight *Marshal*, an officer of the royal household responsible for the observance of the law 'within the king's house and verge', ie within a radius of about 12 miles from the king's palace. The prison closed in 1842, but features in Dickens' *Little Dorrit* (1856) as the grim birthplace of the heroine. Dickens' own father had been imprisoned for debt there in 1824.

A name that can also be placed in this criminal category is that of *Whip-ma Whop-ma Gate*, York. The name probably refers to the whipping post and pillory that stood at the end of the street. It was recorded in 1546 as *Whipnam Whapnamgate*, with the first word apparently representing *whipman*, 'man who wields a whip'.

Modern street names can derive from familiar public buildings. *Bath Row*, Stamford (Lin), is named from former public baths first built here in 1722. *Old Post Office Lane*, Maidenhead (Ber), stands on the site of the earlier Maidenhead Post Office, and *Post Office Court*, off Lombard Street in the City of London, is on the site of London's first General Post Office, opened in 1705. There is still a post office nearby. *Police Street*, Manchester, takes its name from the police office originally opened in King Street in 1772. *Bank Plain* and *Bank Street*, Norwich, are so named for Barclays Bank, built in the late

1920s and now occupying the whole west side of *Bank Plain*. *Bank Street*, Lincoln, similarly, derives from the Lincoln Savings Bank, established here in about 1816.

Museum Street, London, approaches the main entrance of the British Museum, while *Museum Road*, Oxford, is named for the University Museum. *Coptic Street*, London, runs parallel to *Museum Street*, and is also named for the British Museum, with a specific reference to its important collection of manuscripts from the Egyptian or Coptic Christian Church. *Library Place*, Wapping, east London and *Library Street*, near St George's Circus, Southwark, south London, are both named for local public libraries.

Turning from study to sport and entertainment, *Stadium Road*, as its name implies, runs by the stadium of the Garrison Athletic Ground on Woolwich Common, south London. The name is found elsewhere by sports grounds. In Hull (Hum), *Astoria Crescent* was so named in 1935 for the Astoria Cinema, opened the previous year. *Globe Street*, Southwark, south London, was named for the Globe Inn which was itself probably named for the historic Globe Theatre of Shakespeare's day, on Bankside. (For street names in general derived from inns, see Chapter 9, p. 125.) *Playhouse Yard*, in the City of London, has a name that refers to the Blackfriars Playhouse, opened here by the actor Richard Burbage in 1600. *Scala Street*, just west of Tottenham Court Road, London, is named for a theatre built here in 1904 on the site of an earlier theatre dating from 1772. *Strand Close*, Hull, is named for the old Strand Cinema nearby on Beverley Road. The cinema opened in 1914, closed in 1960 and was destroyed by fire in 1968, leaving its memorial only in the reminiscences of its former patrons and in this street name. *Theatre Street*, Norwich, bears a reference to the theatre built here in 1826 on the site of an earlier theatre erected in 1757. The second theatre was itself burned down in 1934 so that the present one, built the following year, is the third on the site. London's *Theatre Street*, in Battersea, south London, takes its name from the Shakespeare Theatre built here on Lavender Hill in about 1910.

Great Northern Terrace, Lincoln, is one of a select number of streets named for a regional railway company. This street crosses the railway, formerly the Great Northern line. In similar manner *Midland Road*, by St Pancras station, London, derives its name from the old Midland Railway Company, whose terminus the station was. *Midland Street*, Hull (Hum), has a name referring indirectly to Paragon station and directly to the York and North Midland Railway, its former owners. *South Western Road*, in Twickenham, southwest London, is not far from St Margaret's station, on the line from Richmond, and is so named for the London and South Western Railway.

Station Approach was mentioned in Chapter 5 as a common directional name. Sometimes the name may be locational, as for *Station Square*, by St Mary Cray station, southeast London, or *Station View*, by Greenford Station, west London. The latter small street runs parallel with *Belle Vue*, and may have been punningly suggested by it (or vice versa). Hull (Hum) has *Station Drive* as a reminder of the former Stepney station which originally served the dock railway and which closed in 1964.

Street names based on the word *bar* often refer to a gate, especially at the entrance to a town. The term in this sense is related to *barrier*, and implies a physical object that needs to be opened or negotiated before entrance can be made. *Above Bar*, Southampton (Ham), is thus named for *Bargate*, the Norman gate that formerly stood at the city entrance, from which the street runs uphill to the north. *Chapel Bar*, Nottingham, similarly takes its name from the *bar* or gate that itself gave the name of the former *Bargate*, where *gate* represents Old Norse *gata*, 'road'. *Bargate*, Newark (Not), still exists as a street that was originally closed by a *bar*. Whereas these streets were named for the gate, however, in York the reverse applied, and the gates were named for the streets. The gates were (and still are) the main entrances to the city, and are four in number: *Bootham Bar* to the north, *Micklegate Bar* to the west, *Monk Bar* to the east (named for *Monkgate*) and *Walmgate Bar* to the south. The same term gave the name of *The Bars*, Chester, referring to a gate taken down in 1770. The original name of the gate was *Foregate*, so called because it stood at the outer end of *Foregate Street* in front of (ie *before*) *Eastgate*. More straightforwardly, *Tollbar Street*, Stockport (GtM), is named for a former tollbar across it. *West Bars*, Chesterfield (Der), is a street named for its location at the western bars or entrance to the New Market, now a car park.

Littlegate Street, Oxford, takes its name from one of the seven medieval gates of the city. *Northgate*, Lincoln, as the street between Eastgate and Church Lane, is fairly obviously named for the northern gate of the city. *Posterngate*, Hull (Hum), takes its name from the small gate that gave entrance to the city here through a medieval wall. *Lower Yott*, Hemel Hempstead (Her), running down to the major St Albans Road, has a name meaning 'lower gate', from a dialect form of Old English *geat*, 'gate'. There may at one time have been a 'Higher Yott', although there is no street so named now.

In the case of *Alvin Street*, Gloucester, the term for gate is no longer present in the name. Even the gate itself no longer exists. As *Alvingate*, its name seems to have meant 'Ælfwine's gate', from the Anglo-Saxon personal name. The gate gave entrance to the city from the north. *The Stonebow*, York, until recently *Stonebow Lane*, is a street that must have been named for a stone arch over it at some time. The name is Old Norse in origin, from *stein*, 'stone', and *bogi*, 'bow', 'arch'. An identically named stone arch stood at one time as the main entrance to the city at Lincoln.

Turl Street, Oxford, takes its name not so much from a conventional gate as from a turnstile or other revolving mechanism. The street runs from Broad Street to the High Street, and at one time had a turnstile at the former end. A more modest street entrance is alluded to in the name of *Cheney Lane*, Oxford, which presumably had a chain at one end, or even at both ends. *Chain Row*, Ilkeston (Der), refers to a row of cottages which had a line of chained posts in front of them before the present road was widened in 1979, when the cottages were demolished.

Burges, Coventry (WMd), has a name meaning 'bridges', referring to its location between St John's bridge and another bridge near the Free School. *Galliard Street*, Sandwich (Ken), is named for the former Galliots Bridge. *Ivegate*, Bradford (WYo), is also named for a bridge, as more transparently is *Stonebridgegate*, Ripon (NYo). (See Chapter 10, p. 132 for the origins of bridge names themselves.)

In concluding this review of streets named for buildings and other manmade objects, mention should be made of *Birdcage Walk*, Westminster, west London, apparently so called from the aviary of James I in St James's Park, of *Telegraph Hill*, at the western end of Hampstead Heath, London, where a telegraph linked the capital with Dover and the Continent in the 18th century, and of *Beacon Hill*, Holloway, north London, where a more basic means of signalling once stood, the light of its fire either warning of danger or telling of victory.

9. NAMES FROM INNS

Hardly a single town in England exists without one or more streets bearing the name of an inn, whether the latter is still in existence or not. An inn's name has traditionally long been depicted by its sign, which not only indicates the location of the inn itself but serves as a 'landmark' for someone unfamiliar with the neighbourhood. It thus also serves to indicate a particular street, and in this respect it may be assumed that where a street is named for an inn, it is named more from its prominent sign than from the building itself.

The oldest inns in England date back to the alehouses of medieval times. A few street names derived from inns thus also date from this period. Most, however, are of later date, when the inn had virtually become a public house or even a hotel. (The earliest citation in the *Oxford English Dictionary* for *public house* in its modern sense is dated 1669.)

Since names in this category form an interesting subgroup in their own right, they are treated separately here, and the uniformity of the names makes for a presentation that is rather different from the discursive text found in most other chapters. The selection of names that follows is therefore arranged in a single alphabetical run for ease of reading, the order of information being: street name, location, earliest recorded date (where known), comments (where appropriate). In many cases the street has adopted the name of the inn (alehouse, tavern, pub, hotel, etc) unchanged, but there are instances where the name has been abbreviated. Unless otherwise indicated, it can be assumed that inn name and street name are identical. Inevitably, as for street names in general, certain inn names repeat. Some towns have or had more than one inn of the same name (see *Bull Lane*, Norwich, for example). It should be borne in mind that London's *Clement's Inn*, *Gray's Inn Road* and *Lincoln's Inn Fields*, together with a few related names, do not refer to an inn in the popular sense but to a hostel for barristers and law students at one of the Inns of Court.

Adam and Eve Court, Westminster, west London, from a former tavern;

Adelaide Grove, Hammersmith, west London, from a pub named for Queen Adelaide (1792-1849);

Alfred Road, Westminster, west London, from the Royal Saxon pub, whose own name alludes to the Saxon king Alfred the Great;

Anchor Street, Lincoln (1868), probably from the Blue Anchor inn;

Angel Place, Southwark, south London;

Angel Road, Edmonton, north London;

Angel Road, Northampton (1504), earlier *Knyghtstrete*;

Angel Row, Nottingham (1587);

Bear Alley, City of London, from an inn recorded here in 1666;

Bear Lane, Oxford (no early records are known);

Beaumont Road, Windsor (Ber) (1910), from the Beaumont Arms inn;

Bedford Road, Clapham, southwest London, from the Bedford Arms pub, itself named for the Duke of Bedford's property in nearby Streatham;

Bedford Street, Norwich (1885), from the Bedford Arms;

Bell Avenue, Norwich (1920), from The Bell hotel, said to have originally been the Blue Bell;

Bell Inn Yard, City of London;

Bell Lane, Spitalfields, east London, from an inn recorded in 1761;

Bell Street, Maidenhead (Ber), from The Bell hotel;

Black Horse Court, Southwark, south London, from an inn recorded in 1799;

Black Jack Street, Cirencester (Glo) (1779), a 'black jack' being a type of tarred leather jug for ale;

Black Lion Lane, Hammersmith, west London, from a late 18th-century inn;

Black Lion Yard, Whitechapel, east London, from an inn recorded in 1668;

Black Swan Yard, Southwark, south London, from an inn recorded in 1826;
Blue Anchor Yard, Whitechapel, east London (18th century);
Blue Boar Lane, Leicester (1716);
Blue Boar Row, Salisbury (Wil) (1751);
Blue Boar Street, Oxford (17th century), from an inn on the site of the present Museum of Oxford, on the corner of St Aldate's; the inn was pulled down in 1893 when the Oxford Public Library was built here;
Boatsman's Close, Ilkeston (Der), from the former Jolly Boatman inn nearby;
Britannia Road, Hammersmith, west London, from a tavern recorded in 1770;
Bull Alley, Southwark, south London, from The Bull in Pound inn once here;
Bull Close, Norwich (1766), see next name;
Bull Lane, Norwich (1845; Norwich had five inns so named at this time);
Bull Street, Birmingham (1667), earlier *Chapelstrete*;
Cardinal Cap Alley, Southwark, south London, from an inn here in 1588;
Castle Court, City of London;
Castle Hill, Maidenhead (Ber), from the Windsor Castle pub;
Castle Road, Camden, north London, from a mid-18th century inn;
Chequer Street, St Albans (Her), from an inn that became the Queen's Hotel;
Cherry Lane, Norwich (1696), from the Cherry Tree inn;
Clarence Road, Hackney, east London, from the Duke of Clarence inn, itself named for George III's second son, later William IV (1765-1837);
Coach and Horses Yard, Westminster, west London, from the former name of the Burlington Arms pub;
Cow Hill, Norwich (1830), from the 18th-century Red Cow inn;
Crane Street, Salisbury (Wil) (1637), from a 15th-century inn;
Crown Street, Harrow, west London;

Dogs Head Street, Ipswich (Suf) (17th century);
Dolphin Court, Petersfield (Ham) (18th century), from a hotel pulled down in 1962;
Dragon Street, Petersfield (Ham) (1736), from a 17th-century inn also known as the Green Dragon;
Eagle Street, Ipswich (Suf), from the Spread Eagle inn;
Eaglet Place, Mile End, east London, from 19th-century Golden Eagle pub;
Elephant and Castle, Newington, south London (18th century), from a famous old coaching inn; its name does not represent Spanish *Infanta de Castile*, as popularly explained, but describes the crest of the Cutlers' Company, depicting an elephant with a howdah in the form of a small castle on its back;
Falcon Close, Southwark, south London, from an inn here from 1541 to 1808;
Fountain Street, Bethnal Green, east London, from a 19th-century inn;
George Inn Yard, Southwark, south London, from a 16th-century (or earlier) coaching inn, the only one of its kind to survive in Southwark;
George Lane, Marlborough (Wil);
George Street, St Albans (Her);
Globe Road, Bethnal Green, east London, from the Old Globe pub, formerly the Globe;
Goat Street, Bermondsey, south London, from a tavern recorded here in 1672;
Golden Ball Street, Norwich (1696);
Golden Dog Lane, Norwich (1696), from an alehouse recorded here in 1626;
Grapes Hill, Norwich (1920), from The Grapes pub;
Great Swan Alley, City of London, from The Swan inn formerly here;
Green Dragon Yard, Whitechapel, east London, from an inn recorded in 1746;

Greyhound Road, Fulham, south London, from an early 19th-century inn;

Half Moon Lane, Dulwich, southeast London;

Half Moon Street, Sherborne (Dor) (1802);

Half Moon Street, Westminster, west London (1710);

Ham Yard, Westminster, west London, from an inn recorded here in the early 18th century, and subsequently renamed The Lyric;

Harrow Lane, Maidenhead (Ber), from The Harrow pub;

Hertford Street, Westminster, west London, from the Hertford Arms, an 18th-century inn, itself named for the Marquis of Hertford;

Hollybush Row, Oxford, from a 17th-century inn on the site of the present Royal Oxford Hotel in Park End Street;

Keyfield, St Albans (Her), from a 15th-century inn later called the Cross Keys or Peter Keys;

Magpie Alley, City of London; an inn of the name was recorded here in 1761, but see this same street in Chapter 6, p. 75;

Magpie Lane, Oxford, from a 17th-century alehouse;

Maidenhead Street, Hertford, from a 17th-century inn; for a consideration of *Maiden* names, see Chapter 3, p. 48;

Mariners Lane, Norwich (1720), from The Three Mariners inn;

Merlin Street, Clerkenwell, north London, from the New Merlin's Cave tavern formerly here;

Mitre Court, City of London, from the Mitre tavern, recorded as being here in 1475 but destroyed in the Great Fire of London (1666);

Nag's Head Passage, Stamford (Lin), from a pub (now the Lincolnshire Poacher);

Neptune Street, Southwark, south London, from a pub here in the 1830s;

North Star Lane, Maidenhead (Ber), from a pub named for a famous locomotive;

Orchard Place, Poplar, east London, from The Orchard House, an inn formerly here, demolished in about 1850;

Paragon Street, Hull (Hum) (19th century);

Plough Road, Battersea, south London, from a 17th-century inn;

Poland Street, Westminster, west London (18th century), from the 17th-century King of Poland inn, so named for the victory over the Turks of John Sobieski, king of Poland, in 1683;

Prince's Street, Norwich (1830), from the Prince's Inn;

Rampant Horse Street, Norwich (1720), from a 13th-century inn, itself named from the horse market; the inn was here until the late 19th century;

Red Lion Court, City of London, from a former inn;

Red Lion Square, Holborn, north London (1691), from a former tavern;

Red Lion Square, Stamford (Lin), from an inn that ceased to exist some time before 1720;

Red Lion Street, Finsbury, north London (1746), earlier called Garden Alley;

Red Lion Street, Norwich (1696), from an inn here until the 19th century;

Rose Lane, Norwich (1688), from a tavern;

Rosemary Lane, Norwich (1720), from a tavern;

Ship Street, Oxford (1772), from an alehouse, whose original name may have been *Sheep* from the sheepmarket formerly nearby;

Star Road, Fulham, south London, from the former Seven Stars inn here;

Sun Lane, Maidenhead (Ber), from a former coaching inn;

Sun Street, Hackney, east London, from an inn recorded here in the mid-17th century;

Sun Street, Hitchin (Her); in the 18th century called *Angel Street*, from an earlier inn;

Swan Lane, Coventry (WMd) (1411);

Swan Lane, Guildford (Sur), from an 18th-century or earlier inn;

Swan Lane, Norwich (1830);

Swan Walk, Chelsea, west London (1794); the inn was demolished when the Chelsea Embankment was built in 1871;

Tabard Street, Southwark, south London (1877), a modern renaming of the former *Kent Street*, leading south to the Old Kent Road; the name refers to the inn made famous by Chaucer as the one where the pilgrims gathered before setting out for Canterbury; nearby streets have Chaucerian names, eg *Manciple Street*, *Pardoner Street*, *Pilgrimage Street*, *Prioress Street*;

Tavern Street, Ipswich (Suf) (1620), for the many taverns or inns here;

Ten Bell Lane, Norwich (1830), from an 18th-century inn, the Ten Bells;

Three Colts Lane, Bethnal Green, east London (1746);

Three King Lane, Norwich (1885), from an inn whose sign represented the Three Wise Men;

Trooper Road, Aldbury, near Tring (Her), from an 18th-century alehouse, renamed The Valiant Trooper in the late 19th century;

Tunsgate, Guildford (Sur), from The Tun, a 16th-century inn;

Upper Goat Lane, Norwich (1766), earlier *Goat Lane* (1696);

Wellington Lane, Norwich (1845), from a 17th-century tavern;

Wellington Lane, Stamford (Lin), from the Wellington Tavern, an early 19th-century pub;

White Hart Lane, Tottenham, northeast London (1600);

White Hart Yard, Southwark, south London, from a 14th-century inn demolished only in 1889;

White Horse Lane, Stepney, east London, from a 19th-century inn here;

White Horse Street, Baldock (Her), earlier called *Broad Street*;

White Lion Street, Norwich (1819);

Woolpack Street, Nottingham (1641).

10. BRIDGE NAMES

In Chapter 5 bridges were considered as strategic objects that gave their names to the roads which led to them. And insofar as bridges are thoroughfares, they themselves rank as streets for the purposes of this book, and so have the present chapter devoted to them and their names.

What sort of names are they?

With the exception of the giant structures that take major roads over rivers like the Humber and Severn, England's most familiar bridges are almost certainly those in London. It is therefore useful to begin by examining their own names.

Working downstream from Kingston upon Thames to the Thames estuary, London has 19 major bridges: *Kingston Bridge*, *Richmond Bridge*, *Kew Bridge*, *Chiswick Bridge*, *Hammersmith Bridge*, *Putney Bridge*, *Wandsworth Bridge*, *Battersea Bridge*, *Albert Bridge*, *Chelsea Bridge*, *Vauxhall Bridge*, *Lambeth Bridge*, *Westminster Bridge*, *Hungerford Bridge*, *Waterloo Bridge*, *Blackfriars Bridge*, *Southwark Bridge*, *London Bridge* and *Tower Bridge*. All these are road bridges except for *Hungerford Bridge*, which is a rail and footway bridge. It will be noticed that almost all are named for the district or locality where they cross the Thames, the four exceptions being *Albert Bridge*, *Hungerford Bridge*, *Waterloo Bridge* and *Tower Bridge*. *London Bridge* is obviously named for the capital itself, and represents the one and only bridge over the Thames here in Roman times. Of the 14 bridges with district names, nine are named for districts on the south bank of the Thames (Kingston, Richmond, Kew, Putney, Wandsworth, Battersea, Vauxhall, Lambeth and Southwark) and five for districts on the north bank (Chiswick, Hammersmith, Chelsea, Westminster and Blackfriars).

If a bridge with a district name is regarded as leading to that district, then it will be seen that the westernmost bridges lead out of what is now Greater London to the south and west, while the easternmost lead into London proper from the south. This produces the general conclusion that the named district is the one that people aim to reach by crossing the particular bridge.

London Bridge is thus the one that takes the traveller into London. Yet although *Southwark Bridge* is named for a district south of the river, it patently serves to take traffic into central London. Indeed, it carries the A3 from the south. So why does its name appear to run counter to the general principle?

To answer this question one needs to examine the history of the various bridges. *London Bridge* has the earliest recorded name, dating from the 10th century (as *Lundene Bricge*), and it represents the oldest bridge of all, the one that was originally built during the Roman occupation of the first four centuries AD. The fact that it is the oldest bridge explains why it bears the name of the capital itself. *Southwark Bridge*, on the other hand, is a modern bridge, the first of the name being built only in 1819 in response to demand for a new bridge between London Bridge and Blackfriars Bridge. In the event it could hardly have been given any name other than Southwark Bridge, although 'City Bridge' would have better preserved the logic of naming a bridge for the place to which it leads.

It might be supposed that *Waterloo Bridge* is named for the district here south of the river. In fact it is the other way round, and the district takes its name from the bridge. The first bridge of the name was built in 1817 and opened by the Prince Regent that year on the second anniversary of the Battle of Waterloo, 18 June. When under construction (from 1811) its planned name was *Strand Bridge*, since it would lead, as it still does, to this major London thoroughfare. But the victory of the British and Prussian forces over the French at Waterloo inspired a change of name, authorised by Act of Parliament in 1816, for 'the said bridge when completed will be a work of great stability and magnificence, and such works are adapted to transmit to posterity the remembrance of great and glorious achievements'. Waterloo Station, opened in 1848, is thus named for the bridge and helped establish the name for the district.

Albert Bridge was opened in 1873 and named, like many other late 19th-century structures, in memory of Prince Albert, Queen Victoria's husband, who had died in 1861. *Tower Bridge*, opened with great ceremony in 1894, is obviously named for the Tower of London, to which it leads. It was long the only bridge to have been built below London Bridge (see *Queen Elizabeth II Bridge*, below, p. 139), and in its relatively short life has become a familiar symbol of both London in particular and Britain in general.

Hungerford Bridge has the most original name of all London's bridges. Hungerford, after all, is not a London district. The present bridge was built in 1864 to carry the South Eastern Railway over the Thames to its new terminus at Charing Cross. It replaced an earlier bridge of 1845 that was built to serve Hungerford Market, on the north bank of the Thames. The original bridge was thus named for the market, which itself took its name from Sir Edward Hungerford, who built it in 1682 on the gardens of his family house here after the latter had burned down in 1669. It was intended to rival Covent Garden, but was soon outclassed. Hungerford Bridge was for many years known by the logical name of *Charing Cross Bridge*, and appears as such regularly on 19th-century maps. However, it is the historic name that has prevailed, despite the fact that Hungerford Market was demolished in 1863 to make way for Charing Cross Station, with only tiny *Hungerford Lane* remaining as its street-level memorial.

Around the country, many towns have a bridge that bears the name of the town itself, especially when it is the oldest or most important. Examples include *Berwick Bridge* (Ntb) over the Tweed (though superseded as the main road bridge by the *Royal Tweed Bridge*, opened in 1928), *Bideford Bridge* (Dev) over the Torridge, *Burton Bridge* (Sta) over the Trent, *Henley Bridge* (Oxf) over the Thames, *Kingston Bridge* (GtL) over the Thames, *Maidenhead Bridge* (Ber) over the Thames, *Maidstone Bridge* (Ken) over the Medway, *Poole Bridge* (Dor) over the Quay Channel (between Holes Bay and Poole Harbour), *Reading Bridge* (Ber) over the Thames, *Rochester Bridge* (Ken) over the Medway, *Windsor Bridge* (Ber) over the Thames (although only a footbridge) and *Workington Bridge* (Cum) over the Derwent.

Bristol Bridge, recorded in the 12th century, was long the city's main road to the south until superseded in that role by the more modern *Temple Bridge*, named for the historic church here (see *Temple Gate*, Chapter 6, p. 84).

In some towns the main bridge is named for the river that it crosses. This might seem to be the obvious name for a road bridge, but it is found less frequently than might be supposed. London, for example, has no 'Thames Bridge'. The reason is a logical one. A bridge is designed to lead from place A to place B (or simply to place A), not to the river itself. It is a manmade structure, leading to a manmade inhabited area, and as such is usually named for that area. Even in a rural region, a bridge is rarely named for the river but instead often has its own distinctive name (see below). Modern major bridges over estuaries all bear river names, however, such as the *Severn Bridge* taking the M4 from Bristol to south Wales and the *Humber Bridge*, carrying the A15 to Hull (Hum).

Towns where the main bridge (or an important bridge) is named for its river include the *Royal Tweed Bridge*, Berwick (Ntb) (mentioned above), *Trent Bridge*, Nottingham, *Trent Bridge*, Burton (Sta) (with *Burton Bridge*, mentioned above, forming *Bridge Street*), *Edenbridge*, Carlisle (Cum), *Old Dee Bridge*, Chester (originally *Dee Bridge* but in recent times called 'old' to be distinguished from the 'new' *Grosvenor Bridge*, see below), *Wye Bridge*, Hereford, *Greta Bridge*, Keswick (Cum), *Ouse Bridge*, York and *Tyne Bridge*, Newcastle (T&W). *Wearmouth Bridge*, Sunderland (T&W), crossing the Wear between the river proper and Sunderland Harbour, also comes into this category. In Exeter (Dev) the two modern road bridges *Exe Bridge North* and *Exe Bridge South* have replaced the original medieval *Exe Bridge*.

The main bridge in some towns is called simply *Town Bridge*, as in Boston (Lin) over the Witham, Christchurch (Dor) over the Avon, Northwich (Che) over the Weaver, Peterborough (Cam) over the Nene, Stamford (Lin) over the Welland and Weymouth (Dor) dividing the Backwater from the

Harbour. Other bridges have self-descriptive names, such as *Great Bridge* (or *Big Bridge*), Tonbridge (Ken), over the Medway ('Town Bridge' here would have effectively duplicated the town's own name), *High Bridge* and *Low Bridge*, Knaresborough (NYo), respectively upstream and downstream over the Nidd, *Long Bridge*, Barnstaple (Dev), over the Taw, as a restored 13th-century bridge (*Barnstaple Bridge*), *North Bridge*, Colchester (Ess), over the Colne, *West Bridge*, Leicester, over the Soar, *West Bridge*, Northampton, over the Nene, and *Nether Bridge*, Kendal (Cum), as the one furthest downstream over the Kent. Shrewsbury (Shr) lies on the Severn less than 10 miles from the Welsh border. Its two main bridges are *English Bridge*, east of the town centre, leading to 'mainland' England, while *Welsh Bridge*, to the west, takes the road to Wales.

Sluice Bridge and *Haven Bridge*, Boston (Lin), cross the lower reach of the Witham. *Haven Bridge*, Yarmouth (Nor), crosses an anchorage on the Yare. Newcastle (T&W) has six bridges over the Tyne, two of which have self-descriptive names: *Swing Bridge* and *High Level Bridge.* The latter, 111 ft (34 m) above the river, carries both road and railway, and was the largest bridge in the world when it was built in 1849.

As in London, some town bridges are named for the districts, suburbs or villages to which they lead. A familiar example is *Clifton Suspension Bridge*, Bristol, over the Avon. Other instances are *Bedminster Bridge* and *Bath Bridge*, Bristol, over the Avon, the latter leading to the *Bath Road* and so to the city of Bath 13 miles away, *Redcliffe Bridge*, Bristol, over the Floating Harbour, *Queen's Park Bridge*, Chester, over the Dee, *Tuckton Bridge*, Christchurch (Dor), over the Stour, *Framwellgate Bridge* and *Elvet Bridge*, Durham, over the Wear, *Myton Bridge*, Hull (Hum), over the Hull, *Skerton Bridge*, Lancaster, over the Lune, *Ludford Bridge*, Ludlow (Shr), over the Teme, *New Redheugh Bridge* (formerly *Redheugh Bridge*), Newcastle (T&W), over the Tyne, *Carrow Bridge*, Norwich, over the Wensum, as the most recent (1923) of the city's many bridges, *Harnham Bridge*, Salisbury (Wil), over the Avon,

Porthill Bridge, Shrewsbury (Shr), over the Severn (as a footbridge), *Clopton Bridge*, Stratford-on-Avon (War), over the Avon, and *Westham Bridge*, Weymouth (Dor), over the northern end of the Backwater. Less commonly, *Bracebridge Bridge*, Lincoln, over the Witham, gave the name of the district to which it leads. Its own name is said to derive from a dialect word *brash*, meaning 'brushwood', 'small branches', although this seems an unlikely origin for a bridge carrying an important Roman road (Ermine Street) over the river here. The bridge thus reduplicates its own name.

As already mentioned in Chapter 5, many streets leading to a bridge take their names from the bridge. In some instances, however, the bridge is named for the street. An example is *Causey Bridge*, Derby, where a street called *The Causey* ('the causeway') crossed the Derwent on top of the weir at a point where the modern *St Alkmund's Way* runs over the river.

Some bridges are fairly obviously named for nearby buildings or structures, existing or historic. Bridges named for former religious houses are fairly common, and include *Blackfriars Bridge* and *Whitefriars Bridge*, Norwich, over the Wensum, *Friary Bridge*, Guildford (Sur), over the Wey, and *Greyfriars Bridge*, Hereford, over the Wye. (London's *Blackfriars Bridge*, opened in 1769, took its name from the district of Blackfriars rather than directly from the medieval Dominican monastery.) Other bridges of this type include *Gaol Bridge*, Derby, over the Derwent, named for the former prison here, *Station Bridge*, Ipswich (Suf), over the Orwell, for the railway station built in the 1840s, *Duke's Palace Bridge*, Norwich, over the Wensum, for the palace of the Dukes of Norfolk formerly here, *Foundry Bridge*, Norwich, for the iron foundry on the west bank, and *Monument Bridge*, Hull (Hum), over the Hull, formerly linking Prince's Dock and Queen's Dock, named for the memorial here to William Wilberforce (now moved to Wilberforce Drive).

Magdalen Bridge, Oxford, over the Thames, is named for Magdalen College, whose tower is a conspicuous landmark at its western (city) end. *Magdalene Bridge*, Cambridge, is similarly named for the college nearby. But whereas Oxford's *Magdalen Bridge* is the only one named for a university college, Cambridge, thanks to its Backs, has several bridges named for the colleges to which they respectively lead over the Cam. They are all footbridges and include *St John's Bridge*, *Trinity Bridge*, *Clare Bridge*, *King's Bridge* and *Queens' Bridge*. *Garret Hostel Bridge*, leading to Trinity College, takes its name from the watchtower or lookout post that gave the name of *Garret Hostel Lane* (see Chapter 8, p. 118). The timber-built *Queens' Bridge* is popularly nicknamed *Mathematical Bridge*. The present bridge, built in 1902, is a copy of the original of 1750, which was constructed to a mathematical design that needed no nails.

Apart from *St John's Bridge*, St John's College has the famous *Bridge of Sighs* linking the 19th-century New Building with the Third Court. It is so named simply because it is covered and in some sense resembles the Bridge of Sighs in Venice. The bridge linking the two parts of Hertford College, Oxford, over New College Lane, bears the same name, although it is hardly a street in the normal sense of the word. A third bridge of the name is that by Northgate, Chester, over the Shropshire Union Canal.

Certain bridges are named more elliptically for buildings or institutions. *St Mary's Bridge*, Derby, over the Derwent, was originally a medieval bridge with a chapel dedicated to St Mary. The chapel is still there. *St Miles Bridge*, Norwich, over the Wensum, takes its name from St Michael's church nearby. The first name *Miles* evolved in some cases as a pet form of *Michael*.

King's Bridge and *Queens' Bridge* are here only indirectly named for royalty. Many towns have bridges with 'royal' names that relate directly to a king, queen or other member of the royal family, usually one reigning or living at the time it was built or one who may actually have opened it. On the other hand, some royal bridge names were given commemoratively, after the death of a monarch or other royal personage. This notably applies to bridges named *Albert Bridge*, alluding to

Prince Albert (1819-61), consort of Queen Victoria. Both this name and that of *Victoria Bridge* itself are among the most widely found. (London's *Albert Bridge* has already been mentioned, above.) Examples of these and other royal names include *Albert Bridge*, Stamford (Lin), as a footbridge (dedicated in 1863) over the Welland, *Victoria Bridge*, Kendal (Cum), over the Kent, *Victoria Bridge*, Manchester, over the Irwell (opened in 1839), *Victoria Bridge*, Stockton-on-Tees (Cle), over the Tees, and *Victoria Bridge*, Leamington (War), over the Leam. It was Queen Victoria who, after visits to the last-named town, granted it the right in 1838 to add the prefix of its full official name, Royal Leamington Spa. Predictably, perhaps, Windsor (Ber) has both *Victoria Bridge* and *Albert Bridge*. They lead over the Thames from the Home Park, where Queen Victoria and her consort Prince Albert are buried. West of Windsor Town Bridge, *Elizabeth Bridge* is a modern bridge with a royal name over the Thames. *King Edward VII Bridge* and *Queen Elizabeth II Bridge*, Newcastle (T&W), over the Tyne, are railway bridges, with the latter popularly known as the 'Metro Bridge'. *Queen Elizabeth II Bridge* linking Dartford (Ken) and Thurrock (Ess) over the Thames, however, is a road bridge. It relieves heavy traffic in the notorious Dartford Tunnel and was opened in 1991 by the Queen in person.

Some bridges bear a surname or aristocratic title, often that of a local resident or family, or of the bridge's designer or builder. *Grosvenor Bridge*, Chester, over the Dee, was built in 1831 and named for the Grosvenor family of Eaton Hall. Its original name was *New Bridge*, by comparison with the 'old' *Dee Bridge* (now actually called *Old Dee Bridge*, see above). *Norfolk Bridge*, Shoreham-by-Sea (WSx), over the Adur, originated as a suspension bridge built in 1832 by the Duke of Norfolk. Bath (Avn) derives the name of its famous 18th-century shop-lined *Pulteney Bridge*, over the Avon, from William Pulteney, Earl of Bath. *Wardle Bridge*, Middlewich (Che), over the Trent and Mersey Canal, appears to represent the surname of a local family.

Archival evidence is needed for bridge names that may have become corrupted over the years, or that have no immediately transparent meaning. One such is *Tenant Bridge*, Derby, over Markeaton Brook (now covered over). It was recorded in 1483 as *Tynnyn bridge*, suggesting that the name is a corruption of *tenon*, so that the original bridge was a wooden one built with mortise and tenons. These would have been an unusual feature in the construction of a bridge, and so suggested its name. *Miller Bridge*, Kendal (Cum), over the Kent, was recorded as *the Mill Bridge* in 1730, so is named from the former millhouse here rather than a surname. *Bishop Bridge*, Norwich, over the Wensum, and the city's only medieval bridge, is so named because it originally belonged to the Bishop of Norwich. *Fye Bridge*, Norwich, over the same river, is said to have a name that means 'five bridge', meaning that it was regarded as five times the width of a normal footbridge. Oxford's famous *Folly Bridge*, running south from St Aldate's over the Thames, is named for a former tower that stood halfway across it and that was known successively as *Bachelor's Tower*, *Friar Bacon's Study* and finally *The Folly*. It was removed in 1779. *Hythe Bridge*, in the same city, takes its name from the former *hythe* or landing-place on the Thames here. The name dates from the early 13th century.

The precise source of *Lady's Bridge*, Sheffield (SYo), over the Don, is uncertain. It does not seem to be a name of religious origin. It may have been given to contrast with a former *Lord's Bridge*, although no bridge of this name is known, despite a record of *Lord's Bridge Close* in the 18th century. *Lendal Bridge*, York, over the Ouse, may take its name from the medieval St Leonard's Hospital, formerly nearby. *Phipps Bridge*, Merton, southwest London, over an arm of the Wandle, was recorded in 1535 as *Pypesbrige*, showing that the name refers to a pipe, or conduit.

Fullbridge, Maldon (Ess), over the Blackwater, was probably originally 'foul bridge', that is, it was muddy. It may be no coincidence that the river's own name has a similar meaning. Its water is clear but its bottom is dark with mud.

Galliards Bridge, earlier *Galliots Bridge*, Sandwich (Ken), apparently derives its name from *galliot*, a type of small boat. *Ive Bridge*, Bradford (WYo), over Bradford Beck, has a name that probably relates to the ivy that once covered it. These two bridges gave the street names *Galliard Street* and *Ivegate*, respectively (see Chapter 8, p. 124).

The bridges that have given the names of present towns or districts are frequently no longer in existence. London's *Knightsbridge*, for example, was at a point roughly where the Albert Gate stands today as an entrance to Hyde Park. It was here that the present street of the name (the old *Great West Road*) crossed the Westbourne. The 'knights' were not the aristocrats of current times but the young men or soldiers who were servants of a baron or lord. They must have used the bridge as a meeting-place. Similarly *Redbridge*, now a northeast borough of London, takes its name from a former bridge over the Roding here. The bridge, which stood where the modern Eastern Avenue (as the A12) crosses the river with a modern bridge, was probably so named since it was built of red brick to carry vehicles, by contrast with the nearby *White Bridge*, which was a wooden footbridge.

Tracing the exact location of former bridges that have given the name of current towns and boroughs in this way is an interesting exercise, but one that is not strictly relevant to the present book unless, like *Knightsbridge*, they have also given the name of a street.

What are of importance, however, are the many bridges that take roads over rivers in rural regions. As in towns, such bridges are not often named for the rivers or streams they cross. Instead they have their own distinctive names, of various origins. A sampling of such names now follows.

Coultershaw Bridge, carrying the A285 road over the Rother between Duncton and Petworth (WSx), takes its name from a portion of land here whose own name means 'Cūthhere's spur', with the possessive *s* of the personal name and Old English *hōh*, 'spur of land', coming to give a misleading *shaw*, 'wood'. A somewhat similar name is that of *Cassio Bridge*, taking the A412 over the Gade (and Grand Union Canal) between Watford and Rickmansworth (Her), which evolved from 'Cæg's spur'. The Anglo-Saxon who lies behind this name also gave that of the adjacent Cassiobury Park. *Mapledurham Bridge*, taking the A3 over the railway south of Petersfield (Ham), is named for a medieval manor in nearby Buriton, its own name meaning 'maple-tree estate'. *Semington Bridge*, taking the A350 over Semington Brook between Melksham and Westbury (Wil), is named as much for the nearby village of Semington as for the stream that it crosses. (The stream is itself named for the village, although the latter's name evolved from the earlier name of the stream, *Semnet.*) *Carter's Bridge*, leading the A141 road over a fenland drainage channel between Chatteris and Doddington (Cam), takes its name from a local family called Carter. *Fossebridge* takes the A429 over the Coln between Gloucester and Northleach (Glo), and is rather unusually named for the road itself, the ancient *Fosse Way* (see Chapter 2, p. 18). A hamlet of the name has now grown up there. On the other hand *Rawthey Bridge*, carrying the A683 over the Rawthey between Sedbergh and Kirkby Stephen (Cum), is patently named for the river it traverses.

Some bridge names are of obvious derivation. *Old Mill Bridge*, over a stream on the A6116 between Islip and Corby (Ntp), is of this type, as is *Great Bridge*, taking the A3057 over the Test between Romsey and Stockbridge (Ham). *Stamford Bridge*, now a village at a crossing of the Derwent on the A116 between York and Great Driffield (Hum), is a similar self-descriptive name, albeit of Old English origin. It derives from the stone-paved ford originally here that was replaced early by a bridge (recorded here in the 11th century). The crossing was an important one, taking the Roman road over the Derwent east of York, and was famous as the site of the strategic battle of 1066 in which King Harold of England defeated his brother Tostig and King Harald Hardrada of Norway. The replacement of a ford by a bridge as a river crossing in this way is a logical development, and the evolution is frequently preserved in place-names like this. A similar example is that of *Fordingbridge* (Ham), now a town at the point where the modern A33 road crosses the Avon. As a bridge name, this can be interpreted as 'bridge of the people who lived by the ford'.

Yet another example is *Ferrybridge*, the fine 18th-century bridge that takes the A1 over the River Aire east of Pontefract (WYo). Its name means what it says, so that the bridge replaces a ferry or boat crossing here, entered as *Ferie* in the *Domesday Book* of 1086. Records of the name show that the bridge must have replaced the ferry by the end of the 12th century. *Pontefract* itself means 'broken bridge', and the name of the town is believed to refer to a bridge on the site of the present *Bubwith Bridge*, over the small stream known as the Wash Dike. Its own name derives from that of a local family, who themselves presumably came from *Bubwith* (Hum), a village on the Derwent some 20 miles northeast of Pontefract.

Finally, no consideration of bridges and their names is complete without a mention of *Ironbridge* (Shr), now a town on the side of a gorge in the River Severn. It takes its name from the bridge that was the first cold blast iron bridge in the world, cast in 1778. The bridge is still used by foot passengers to cross the Severn, and itself thus has a 'way name' that validates it for a record here.

11. NAMES OF PERSONAL ORIGIN

The preceding chapters have included several examples of streets bearing personal names, usually as the surname of a person connected with a particular street. There are a large number of street names of this kind, which broadly divide into two types.

First there are the 'landowner' names, including those of medieval or later lords of the manor and the like. The name may itself be not even necessarily the landowner's own but transferred from that of his residence, estate or village. Since such landowners were mostly aristocrats who held several properties, they were able to draw on their various names to give to the different streets that were built on their land. As well as their family names, they would also have one or more noble titles (mostly derived from place-names) to add to their potential street name stock. Such landowners are often, by implication, not even local, but live in a 'seat' some distance from the land where the street is laid out and named for them. Several of them, on the other hand, frequently had a town house in the street that now bears their name, especially in London, so that they were both absent as landowner and present as resident. In the case of episcopal ownership, the name is often that of the bishop's see or diocese, although it may also be adopted from an ecclesiastical title. When land is owned by a college, as is sometimes the case, the street name is usually that of a college living, since no family name is normally involved.

Then there are the local names, more those of people 'on the spot'. These include builders, developers, landlords, noted townspeople such as mayors and council officials, and people closely associated with the street in some way. (Commemorative names given for royal personages, military heroes and the like do not belong here but are considered in Chapter 12, p. 167.)

As may be appreciated, the dividing line between a 'landowner' name and a local one is often quite fine. In some instances it can be proved fairly conclusively that a street is named for a local family, but it may be difficult to establish whether the family were actually landowners or not, or even to say why one family rather than another should have had a particular street named for them.

It is frequently the case that the more recent the name, the better documented will be its history. This means that the converse is true, and that streets having medieval personal names may lack archival detail. Even so, despite the uncertainty in many cases, it is usually possible to identify accurately the name of the person that has passed to the street. Invariably, it is the names associated with the most important cities that are the most reliably documented. This is particularly true of London streets, which are consequently well represented here.

What follows in this chapter is a selection of street names of personal origin of both types. The 'landowner' names are considered first. The multiple names associated with prodigious landowners, such as the Grosvenor and Gunter families, or the Dukes of Bedford and Marquises of Northampton, are not fully enumerated, but there should be enough examples to suggest the manner in which streets were named from a wealthy landowner, from members of his family, and from his various properties. Only one of the many streets named for the less familiar Powell-Cotton family is quoted, and it heads the selection.

Acol Road, Camden, north London was built by the Powell-Cotton family of Quex Park, Birchington (Ken), now a western district of Margate, and given the name of the nearby village of Acol. (The unusual name means 'oak wood', from a combination of Old English *āc*, 'oak' and *holt*, 'wood'.) *Albemarle Street*, Westminster, west London was built in the late 17th century on the site of Clarendon House, which the Duke of Albemarle had bought from Edward Hyde, Earl of Clarendon. He sold it to Sir Thomas Bond, who laid out the street, and who himself gave the name of *Bond Street* (see below). *Alderney Street*, Westminster, west London was originally known as *Stanley Street*, from the landowner. As there were many streets of this name, it was later renamed *Alderley Street* (*sic*), from the branch of the Stanley family whose estate was at Alderley Park (Che). They objected to the use of the name, however, so it was modified to *Alderney Street*, falsely suggesting a connection with the Channel Islands. *Aldford Street*, also in Westminster, was named for property in Cheshire belonging to the landowners here, the wealthy Grosvenor family, Dukes of Westminster. Aldford itself is a village south of Chester.

Ashby Street, Clerkenwell, east London, was built on land owned by the Earl of Northampton, who had a seat at Castle Ashby, near Northampton. *Bainton Road*, Oxford, was built on land owned by St John's College, who held the living of Bainton, near Great Driffield (Hum). Many streets in this district of north Oxford, some 18 in number, are named for St John's College livings or property. (For a full list, see Chapter 13, p. 181.) *Barkston Gardens*, Kensington, west London, was built on land owned by the Gunter family, with Sir Robert Gunter (1831-1905) being the Member of Parliament for Barkston Ash (NYo). The family name also lies behind that of *Gunter Grove*, Kensington. *Battle Bridge Lane*, Southwark, south London, was built on land owned by the abbots of Battle Abbey, who had a house here until the Dissolution of the Monasteries in about 1539.

Beaumont Fee, Lincoln, takes its name from the medieval manor held by the Beaumont family until 1507, *fee* being a legal term for a heritable estate. *Bedford Square*, Camden, north London, was built on the estate of the Earls (later Dukes) of Bedford. (See also *Cardington Street*, *Great Russell Street*, *Ridgmount Street*, *Southampton Row*, *Tavistock Square*, *Woburn Square*, below.) *Belgrave Square*, Westminster, west London, like *Aldford Street* already mentioned, was built on land belonging to the Grosvenor family and is named for another of their Cheshire properties. The *Belgrave* in question is the hamlet south of Chester near the family seat at Eaton Hall, and not, as sometimes stated, the Belgrave that is now a northern district of Leicester. It was bought by Richard Grosvenor, first Earl Grosvenor, in 1758, together with the manor of Eccleston, which in turn gave the name of *Eccleston Square*. Eaton Hall itself gave the name of London's *Eaton Square*, and *Belgrave*

Square produced the name of *Belgravia*, the fashionable district in which all these streets are located.

Bentinck Street, Marylebone, northwest London, is one of the many names here that originate from the family and properties of the Holles family. For its origin, and those of *Carburton Street*, *Cavendish Square*, *Clipstone Street*, *Devonshire Street*, *Great Portland Street*, *Harley Street*, *Henrietta Place*, *Holles Street*, *Margaret Street*, *Mortimer Street*, *Oxford Street*, *Portland Place*, *Welbeck Street*, *Weymouth Street*, *Wigmore Street* and *Wimpole Street*, see Chapter 13. (For *Oxford Street*, see also Chapter 5, p. 62.)

Bickenhall Street, Westminster, west London, was built on the estate of the Portman family, who also owned property at Bickenhall, near Taunton (Som). *Bishop Lane*, Hull (Hum), was built on land owned by the Archbishops of York. *Blandford Street*, Westminster, was also built on the Portman estate. The family had a seat at Bryanston (now a public school), near the Dorset town of Blandford Forum. The house itself gave the name of nearby *Bryanston Square*. *Bolingbroke Grove*, Battersea, south London, takes its name from Henry St John, Viscount Bolingbroke (1678-1751), lord of the manor of Battersea.

Bond End, Knaresborough (NYo), is one of a number of 'landowner' names in the north of England derived from the owner's occupational status rather than his personal name. The origin lies in Old Norse *bondi*, 'peasant landowner', a word that in fact came to be used as a personal name in its own right (and that ultimately produced the English surname *Bond*). Other streets of this derivation include *Bondgate*, Pontefract (WYo), *Bonegate Green*, Ripon (NYo) and *Bonegate Road*, Brighouse (WYo). *Boongate*, Peterborough (Cam), may also have this origin.

Bouverie Street, in the City of London, was built in the late 18th century and is probably named for William Bouverie, Earl of Radnor. The family later became Pleydell-Bouverie and gave the name of *Pleydell Street*, also in the City. *Bracknell Gardens*, Hampstead, northwest London, was laid out on land owned by the Maryon-Wilson family, who owned property at Bracknell (Ber). *Brigg Street*, Norwich, takes its name from Augustine Briggs, mayor and alderman, who owned land here in the 17th century. He is one of a number of examples of landowners who were also local dignitaries. *Brownlow Street*, Stamford (Lin), was laid out in 1840 on the estate of the first Earl Brownlow. His family home was at Belton House, near Grantham (Lin), which gave the name of the adjacent *Belton Street*, laid out at the same time.

Buckingham Street, off the Strand, London, takes its name from the Duke of Buckingham, who had a house here. He sold it in the 1670s on condition that streets built on the land bear his full name and title. This accordingly produced *George Court*, *Villiers Street*, *Duke Street* (now gone), *Of Alley* (now *York Place*) and the street mentioned. (This is not the same Duke of Buckingham as the one who gave the name of Buckingham Palace.) Henry, Humphrey and Edward Stafford, father, son and grandson, Dukes of Buckingham from 1402 to 1521, gave the names of *Buckingham Road* and *Stafford Road*, Petersfield (Ham), in their capacity as lords of Petersfield. The streets are modern, however.

Cadogan Square, Chelsea, southwest London, was laid out in the second half of the 18th century on land leased from Charles, Lord Cadogan (1691-1776). Cadogan's father-in-law was Sir Hans Sloane, whose surname lies behind the fashionable *Sloane Square* and whose first name gave that of *Hans Crescent*, both also in Chelsea. *Cantelupe Road*, Ilkeston (Der), is named for the Cantelupe family, lords of Ilkeston from the 13th century. *Cardington Street* and *Goldington Crescent*, Camden, north London, were built on land owned by the Dukes of Bedford, who held property at the named places: Cardington is near Bedford and Goldington is now a district of Bedford. The adjacent *Chenies Street* and *Eversholt Street* were built on land owned by the same family. Their names derive respectively from the Duke's title as Baron Russell of Chenies, a village near Amersham (Buc), and Eversholt, a village near the family seat at

Woburn Abbey (Bed). *Cheyne Walk*, Chelsea, bears the name of William Cheyne, second Viscount Newhaven, who laid it out in the early 18th century.

Chicheley Street, Lambeth, south London, was built on land owned by the Archbishops of Canterbury, and is named for one of them, Henry Chichele or Chicheley (1362-1443). The street is not far from Lambeth Palace, London residence of the Archbishops of Canterbury. Another episcopal name is that of *Chichester Rents*, off Chancery Lane in the City of Westminster, which at one time led to the London palace of the Bishop of Chichester, Ralph Neville (died 1244). The street was built up with properties available for rent. *College Road*, Willesden, northwest London, was laid out on land that belonged to All Souls College, Oxford, and the college name appears explicitly in that of nearby *All Souls Avenue*. A short distance to the north lies *Chichele Road*, named for the Archbishop just mentioned, who was the founder of All Souls College. *Cornwall Road*, Lambeth, was built on a detached portion of the manor of Kennington, which belongs to the Duchy of Cornwall. *Curwen Street*, Workington (Cum), was named from the Curwen family, lords of the manor of Workington. The Curwens also owned Belle Isle, the island in Lake Windermere, and this gave the name of *Belle Isle Street*, Workington.

The world-famous *Downing Street*, off Whitehall, was built in the late 17th century on land owned by Sir George Downing. *Easton Street*, Clerkenwell, north London, was built on land owned by the first Marquis of Northampton, who held property at Easton Maudit, near Wellingborough (Ntp). *Ely Place*, Holborn, north London, was laid out in 1722 on the site of the London palace of the Bishops of Ely. The bishops had been forced to cede part of their estate to Sir Christopher Hatton in 1576, and his name gave that of the well-known *Hatton Garden* here. *Erconwald Street* in Hammersmith, west London, is another episcopal name, on this occasion from the bishop himself, Erconwald or Erkenwald, Bishop of London in the late 7th century. The street is in the manor of Hammersmith which was held by the Bishops of London. *Euston Road*, Islington, north London, now well known for its railway terminus, was built in the mid-18th century over land owned by the Duke of Grafton, whose seat was at Euston Hall (Suf), south of Thetford (Nor). The Duke's title gave the name of *Grafton Street*, Westminster, and the family name, Fitzroy, that of *Fitzroy Square*, Euston.

Great Ducie Street, Manchester, was built on the estate of the wealthy Ducie family, of Woodchester Park near Stroud (Glo). *Great Russell Street* and *Russell Square*, Camden, north London, bear the family name of the Dukes of Bedford, already mentioned (see *Bedford Square*, above). *Great Winchester Street*, Westminster, west London, takes its name from Lord Winchester, to whom an Augustinian monastery here was left by his father, Sir William Powlet, Lord Treasurer, after the Dissolution of the Monasteries in about 1539. Lord Winchester called it 'Winchester House'. *Green Street*, Cambridge, is named for Oliver Green, of Gonville and Caius College, on whose estate it was built in 1614.

Jermyn Street and *St Alban's Street*, Westminster, west London, were built on land granted by Charles II in 1661 to Henry Jermyn, Earl of St Albans. *Leicester Square*, Westminster, was named for Robert Sidney, second Earl of Leicester, who bought land here in the first half of the 17th century and who built Leicester House on the north side of the square. It was demolished in the late 18th century and its site is now taken up by *Lisle Street*, named for Philip, Viscount Lisle, who had succeeded to the earldom of Leicester in 1677. *Marlborough Road*, St Albans (Her), takes its name from the almshouses built here in 1736 by the Duchess of Marlborough. *Melbury Road*, Kensington, was laid out on land belonging to Lord Ilchester, whose country house was at Melbury (Dor).

Norfolk Street, Sheffield (SYo), is named for the landowners here, the Dukes of Norfolk. *Northampton Road*, Clerkenwell, north London, was built at the turn of the 19th century on land belonging to the Marquis of Northampton. The Marquis had a seat at Compton Wynyates, near Tysoe (Ntp), and this gave the names of *Tysoe Street* and *Wynyatt Street* here. The Marquis was born at Yardley Hastings, near Olney (Ntp), and this gave the further name of *Yardley Street. Onslow Gardens*, Kensington, west London, was laid out in the mid-19th century on land owned by the Earl of Onslow. *Portman Square*, Westminster, west London, is named for the Portman family, to whom the manor of Lilestone in Marylebone had passed in the mid-16th century. The family came from Orchard Portman, near Taunton (Som), and Henry William Portman laid out the square in about 1760. His grandmother belonged to the Seymour family, which in turn gave the name of *Seymour Street* in the vicinity. *Orchard Street* nearby is named for the family's Somerset estate, and the county name gave that of *Somerset Street* here. *Ridgmount Street*, Camden, north London, was built on land owned by the Dukes of Bedford (see *Bedford Square*, above), who held property at Ridgmont, near Woburn Park (Bed). *Rochester Row* in the City of Westminster, London, was at one time a private road belonging to the Dean and Chapter of Westminster Abbey. It is so named because many of the Deans were also Bishops of Rochester. *Rutland Gate*, also in Westminster, takes its name from the Dukes of Rutland, on whose land here it was built. The Dukes had a house here in the 18th century.

Salisbury Square, in the City of London, was built on land that had belonged to the Bishops of Salisbury before the Reformation. Their London house of Salisbury Court was here. *Southampton Row*, Camden, north London, is named for the Earls of Southampton, whose town house stood near its southern end. Rachel Wriothesley, daughter of the fourth Earl, married Lord William Russell, and it was through this marriage that the Russell family, Earls (later Dukes) of Bedford, gained property here in Bloomsbury. Hence the contiguity, round the British Museum, of *Bedford Square*, *Russell Square* and *Southampton Row. Stanhope Gardens*, Kensington, south London, bears the family name of the Earls of Harrington. The first Lord Stanhope of Harrington held the adjacent manor of

Chelsea in the 16th century, and the family's noble title gave, through marriage, the names of *Harrington Square* and *Harrington Street*, Camden, north London. The title itself came from Harrington, near Kettering (Ntp).

Tavistock Square, Camden, north London, was part of the estate of the Dukes of Bedford (see *Bedford Square*, above), and is named from the family's estate at Tavistock (Dev). The title of Marquess of Tavistock has long been that of the Duke's heir. (It is currently borne by Henry Robin Ian Russell, born 1940, son of the 13th Duke of Bedford.) *Westmoreland Buildings*, in the City of London, stands on the site of the town house of the Earls of Westmorland (*sic*). *Winchester Road*, Camden, north London, arose on the estate of Eton College, whose first provost was William Waynflete, Bishop of Winchester. *Woburn Square*, Camden, north London, was part of the estate of the Dukes of Bedford (see *Bedford Square*, above) and takes its name from the family's principal seat, Woburn Abbey, near Leighton Buzzard (Bed). *York Road*, Battersea, south London, is named for the former house here bequeathed to the Archbishops of York in the 15th century.

Many of the streets named above are famous almost as much from their location as from the people whose names, titles and properties lie behind them. The selection below is of names of a much more 'local' type, from medieval persons about whom little is known to active members of a town's community in more modern times. Some of the names, however, are borne by people who became famous in their own right, such as the actor Edward Alleyn, the aviatrix Amy Johnson and the antiquary Elias Ashmole.

Abbey Street, Hull (Hum), despite appearances, is so named for Alderman Thomas Abbey, a member of the local Board of Health, responsible for streets. More famously, *Adam Street*, Westminster, west London, is named for the architect brothers John and Robert Adam, who designed the *Adelphi Terrace* here. The latter name is based on the Greek word for 'brothers', *adelphoi*. *Addison Road*, Kensington, west London, is named for the writer Joseph Addison, who lived at Holland House nearby after his marriage to Lady Holland in 1716. *Holland Road* here has the same connection. *Alfred Street*, Hull, is named for the son of Thomas English, a local shipbuilder who developed this and *English Street*, among others. Hull's *Alfred Gelder Street* is named for a mayor of the city who was a prominent architect and local Member of Parliament at the turn of the 20th century. *Alleyn Park*, Dulwich, southeast London, is named for the famous Elizabethan actor-manager Edward Alleyn, who bought the manor of Dulwich and founded Alleyn's School and Dulwich College here in 1619. *Burbage Road*, Dulwich, is named for Alleyn's main rival, Richard Burbage.

Amanda Close, Hull (Hum), was so named in 1971 for the three-year-old daughter of R. H. Beal, director of the company that built it. In Ilkeston (Der) *Amilda Avenue*, *Margaret Avenue* and *Vincent Avenue* were built by Albert Beards and named for his wife Margaret Imelda and son Vincent. *Amy Johnson Court*, Hull, was named for the famous aviatrix, who was born in Hull. *Ann Watson Street* in the same city bears the name of an 18th-century benefactress whose almshouses stood here. *Applegarth Road*, also in Hull, takes its name from Robert Applegarth, a local councillor and one of the founders of the Trades Union Congress. Another local councillor, George Archer, gave the name of *Archer Street*, Ilkeston. The street was laid out at the turn of the 20th century. More obliquely, *Aristotle Road*, Clapham, south London, was so named in the late 19th century by the classics-loving owner of the land here, a Mr Foster. He also named nearby *Plato Road*. *Arne Street*, Westminster, west London, is named for Thomas Arne, the 18th-century composer, who lived here. *Ashmole Street*, Lambeth, south London, bears the name of the famous antiquary, Elias Ashmole, who lived here in the late 17th century. His name is more generally familiar from Oxford's Ashmolean Museum.

Atkinson Street, Manchester, is named for the wealthy Atkinson family of Manchester. *Aubert Park*, Highbury, north

London, takes its name from the 18th-century astronomer Alexander Aubert, who set up an observatory here. *Babington Lane*, Derby, was laid out in 1789 and named from the Babington family who lived locally. *Bagleys Lane*, Fulham, west London, is named for Robert Bagley, who bought a farm here for use as a market garden in 1812. *Bakewell Road*, Loughborough (Lei), preserves the name of Robert Bakewell (1725-95), a local cattle breeder, who lived at nearby Dishley Grange. *Barchard Street*, Wandsworth, southwest London, commemorates the family of Elizabeth Barchard, a local benefactress, and Robert Barchard, Deputy Lieutenant of Surrey, the county in which Wandsworth was located at the time of his death in 1848. *Barn Street*, Marlborough (Wil), is named for the family of Roger Baron, known to have been living here in the 13th century. *Baylis Road*, Lambeth, southeast London, honours Lilian Baylis, manager of the Old Vic theatre here. Her aunt, Emma Cons, gave the name of nearby *Cons Street*.

Beak Street, Westminster, west London, was built on land acquired here in the 1670s by Thomas Beak. *Bean Street*, Hull (Hum), is named for a 19th-century brewer, Alderman Robert Bean. *Beau Street*, Bath (Avn), built in the 18th century, preserves the memory of Richard 'Beau' Nash (1674-1762), the social arbiter who made Bath a fashionable watering place. Nearby *Nash Street* is also named for him. (He should not be confused with John Nash, the equally famous architect who laid out Regent's Park, London and who designed *Regent Street* there.) *Beavor Lane*, Hammersmith, west London, was so named for Samuel Beavor, who in the 18th century owned a house here called Beavor Lodge. (The name puns on *lodge* as a term for a beaver's den.) Unexpectedly, *Bedding Lane*, Norwich, preserves the name of a family called Badding, known locally in the 13th century. *Bedward Lane*, Chester, somewhat similarly bears the name of Charles Bedward, who built a row of cottages here in the first half of the 19th century. *Beeton Street*, Hull is a memorial to an eccentric property developer in the 19th century, James Beeton. Adjacent streets bear the married surnames of his daughter, sisters and nieces: *Paisley Street*, *Pulman Street*, *Longden Street*, *Evans Square* and *Cookes Terrace*. The district was colloquially known locally as *Beetonville*.

Beetwell Street, Chesterfield (Der), is named for a well or spring owned by the medieval family of *Bete*. *Birkbeck Road*, Hackney, east London, was directly named for Birkbeck School here, and so indirectly for its founder, George Birkbeck (1776-1841), a local doctor. London's famous *Bond Street* was named in honour of Sir Thomas Bond, a financier in favour at court in the 17th century. The streets are now *Old Bond Street* (built first) and *New Bond Street* (its completion in 1721), but the Underground station here bears the original name, without a prefix. *Brown Street*, Manchester, bears the name of Joshua Brown, a wealthy cloth dyer here in the 17th century. The street of the same name in Salisbury (Wil) is recorded in the 13th century (as *Brounstrete*) but may not derive from a personal name. The short parallel streets *Browning Close*, *Robert Close* and *Elizabeth Close*, in Paddington, west London, are named for the poet Robert Browning and his wife Elizabeth. Browning lived near Little Venice here after her death in 1861. *Brushfield Street*, Tower Hamlets, east London, preserves the name of Thomas Brushfield, local representative of the Metropolitan Board of Works for 10 years from 1865. *Burton Street*, Loughborough (Lei), commemorates Thomas Burton, Loughborough's great benefactor in the 15th century.

Calvert Street, Norwich, has a name first recorded only in 1819. It is said to derive from a local councillor. Its medieval name was *Snailgate*, probably not for the mollusc but from a personal name. *Cameron Avenue*, Abingdon (Oxf), pays tribute to Group Captain Neil Cameron, commanding officer of the local RAF station in the early 1960s. Another military name is that of *Captain French Lane*, Kendal (Cum), commemorating a former Parliamentarian officer. *Carline Road*, Lincoln, bears the name of the local Carline family, here in the 19th century. *Caroline Court*, Ilkeston (Der), was given the name of the granddaughter of a local chartered surveyor, Stanley Healey. His grandson gave the name of *Charles Close* and his son that of *Julian Close*. He himself does not appear to be represented, perhaps partly because *Stanley Close* already preserved the name of the Stanley family who were Earls of Derby.

Carroll Place, Hull (Hum), contains a rather remote reference to the famous author of *Alice in Wonderland*, Lewis Carroll, whose maternal grandfather, Charles Lutwidge, lived in Charlotte Street nearby. *Castelnau*, Barnes, southwest London, was built over land belonging to (and named for) a French family who came to Barnes to escape religious persecution at the beginning of the 19th century. The name is unusual in having no generic word (such as *Street*). *Champion Hill*, Camberwell, southeast London, has preserved the name of Sir Claude Champion de Crespigny, who developed the family estate in Camberwell in the second half of the 19th century. A much earlier name is that of *Clasketgate*, Lincoln. It means 'Clacc's gate', from the Old Danish personal name *Klak*, who must have been associated with the place in some way. The name is first recorded in the 13th century. *Charlotte Street*, Bath (Avn), built in 1839, takes its name from Lady Charlotte Rivers, a member of a wealthy local family whose surname is preserved in *Rivers Street*, laid out in about 1770. *Colegate*, Norwich, is a name dating from the same period, and derives either from the Anglo-Saxon personal name *Cola* or a family name *Cole*. *Colet Gardens*, Hammersmith, west London, is named for John Colet, Dean of St Paul's, who founded St Paul's School in 1510. The school moved to Hammersmith in 1880, hence the relevance of the name here.

Colmore Row, Birmingham, is named for a local family, as is *Cooper Street*, Manchester. *Coram Street*, Camden, north London, commemorates Captain Thomas Coram, who established a foundling hospital here in 1741. Coram's Fields stands on the site of the former hospital, which moved to Berkhamsted (Her) in 1926. *Coventry Street*, London, in the City of Westminster, bears the name of Henry Coventry, Secretary of State to Charles II in the 1670s, whose house was here. *Cramer*

Street, also in Westminster, is named for the 18th-century German-born violinist Wilhelm Cramer, who came to settle in London on the invitation of J. C. Bach. A literary association gave the names of *Cranford Avenue* and *Gaskell Avenue*, Knutsford (Che), since the novelist Elizabeth Gaskell, though born in London, was brought up by her aunt in Knutsford and her novel *Cranford* is said to be set in the town. *Crawthorne Road*, Peterborough (Cam), takes its name from a family who lived in the city in the 17th century. *Crisp Road*, Hammersmith, west London, is named for Sir Nicholas Crisp, merchant and royalist, who had a house here in the first half of the 17th century. The 19th-century artist George Cruikshank gave the name of *Cruikshank Street*, Islington, north London. He lived in neighbouring Amwell Street.

D'Arblay Street, Westminster, west London, was so named in 1909 for the writer Fanny Burney, Madame D'Arblay, who lived in Poland Street nearby. The earlier name of the street was *Portland Street*, for the second Duke of Portland (see *Bentinck Street*, above). *Davenant Road*, Oxford, bears the name of Sir William Davenant, who became Poet Laureate in 1638. *Davey Place*, Norwich, is named for Alderman Jonathan Davey, who was responsible for improvements to nearby streets in 1813. *Davygate*, York, takes its name from David le Lardiner, who in the 12th century held the title of Royal Lardiner of the Forest of Galtres. A *lardiner* was a clerk in charge of a larder in a royal or other kitchen. *Dawes Road*, Southwark, south London, has the name of James Arthur Dawes, first mayor of the new Borough of Southwark in 1899. *Deacon Street*, Peterborough (Cam), looks like a name derived from a religious office. It in fact refers to Thomas Deacon, a local woolcomber, who founded Deacon's School here in 1721. A religious office does lie behind *Dean Farrar Street*, Westminster, where Frederick William Farrar lived as rector of St Margaret's, Westminster before he became Dean of Canterbury in 1895.

The novelist Daniel Defoe lived in what is now *Defoe Road*, Stoke Newington, north London, and wrote *Robinson Crusoe* here. *Dover Street*, Mayfair, west London, was built by Lord Dover in the 17th century. *Drakes Circus*, Plymouth (Dev), is a modern name given in honour of the famous navigator Sir Francis Drake, who returned to Plymouth in 1580 after sailing round the world, the first Englishman to do so. *Circus* may be a deliberate punning allusion to his circumnavigation. *Drury Lane*, Westminster, west London, formerly famous for its theatres, takes its name from Sir William Drury, who built his house at the southern end in the 16th century. *Dryden Street*, a small street nearby, is named for the 17th-century poet and playwright John Dryden, whose plays were produced at the Theatre Royal here.

Fearon Street, Loughborough (Lei), is named for a local rector, Archdeacon Fearon, who died in 1885. *Ferensway*, Hull (Hum), takes its name from the city's great benefactor, Thomas Robinson Ferens (1847-1930), who also gave the name of *Ferens Avenue*. A name considered for *Ferensway* when it was laid out as a major road across Hull in 1931 was *Quality Street*. The philanthropist had recently died, however, so his name was adopted as more meaningful and specific instead. *Fletcher Way*, Hemel Hempstead (Her), is not named for a medieval arrow-maker but for a 20th-century dignitary, Alderman Fletcher. *Flood Street*, Chelsea, west London, leads down from the King's Road to Chelsea Embankment, by the Thames. Its name does not refer to an inundation, however, but was given for L.T. Flood, a Victorian benefactor to the parish of St Luke, Chelsea. *Frewin Court*, off Cornmarket, Oxford, commemorates Richard Frewin, Professor of Ancient History, who lived here in the 18th century. The narrow street leads to Frewin Hall, now part of Brasenose College and also named for the professor. *Furmage Street*, Wandsworth, southwest London, is named for William Furmage Palmer, on whose land it was built in 1898.

Gandy Street, Exeter (Dev), is named for Henry Gandy, twice Mayor of Exeter in the 17th century. London's *Garrick Street*, Covent Garden, takes its name from the famous 18th-century actor-manager David Garrick, who lived in the area in the 18th century. *Gay Street*, Bath (Avn), took the name of a surgeon, a Mr Gay, who owned land here at the time when the street was built in 1734. The name happens to be appropriate for a city formerly famous for its 'gay society' (in the original sense of the word). *Gibbon Walk*, Putney, south-west London, commemorates the historian Edward Gibbon, who was brought up in Putney. *Gladstone Street*, Peterborough (Cam), bears tribute not to the famous prime minister but to his brother Thomas, who contested the election of 1841 for the city. *Goodge Street*, St Pancras, north London, is named for John Goodge, a carpenter who had acquired land here in the early 18th century. *Goodramgate*, York, means 'Guthrum's street'. The name is an anglicised form of the Viking personal name *Guthormr*, literally 'god serpent', familiar from the Danish leader and king of East Anglia who fought Alfred the Great in the 9th century. *Grantham Street*, Lincoln, takes its name from a well-known local family, whose own name doubtless came from Grantham (Lin), only 25 miles to the south. *Great Pulteney Street*, Bath (Avn), bears the name of Sir William Pulteney, for whom Robert Adam built *Pulteney Bridge* here (see Chapter 10, p. 139) in 1774. *Laura Place* nearby is named for Pulteney's daughter. *Gresham Street*, in the City of London, honours Sir Thomas Gresham, founder of the Royal Exchange in 1570 and of Gresham College in 1579.

Handel Street, Camden, north London, preserves the memory of the great composer, who was organist to the foundling hospital founded by Thomas Coram (see *Coram Street*, above) on which the street was built. *Handel Way*, Edgware, north London, recalls the composer's association with the Duke of Chandos, his patron, whose country house of Canons (now Canons Park) was here. The Duke himself is commemorated in nearby *Chandos Crescent*, as well as more famously in *Chandos Street*, in the City of Westminster. *Hankey Street*, Peterborough (Cam), preserves the name of Thomson Hankey, local Member of Parliament and a director of the Bank of England in the mid-19th century. *Hardman Street*,

Manchester, takes its name from William Hardman, a successful property developer here in the mid-18th century. *Haverstock Hill*, Hampstead, northwest London, is unfortunately still an obscure name. It dates from at least 1721 and may represent the name of a family who came from what is now the village of Stock (formerly Haverstock), near Billericay (Ess).

Herbert Street, Kentish Town, north London, was built in the mid-19th century by Vincent Herbert. *Herrick Street*, in the City of Westminster, London, is a reminder of the 17th-century poet and clergyman Robert Herrick, who lived nearby. *Hertford Street*, Coventry (WMd), was laid out in the early 19th century and named for the Marquis of Hertford, who built Coventry Park. *Hewlett Road*, Cheltenham (Glo), is named for a local family, represented by William Howlett (*sic*) in 1689.

Hickling Court, Loughborough (Lei), recalls a 17th-century benefactor, William Hickling. *Hillaries Road*, Birmingham, apparently refers to a local family named Hillary. The artist William Hogarth lived in what is now *Hogarth Lane*, Chiswick, west London, in the 18th century, and another artist, Hans Holbein, painted the family of Sir Thomas More at his house in Chelsea in 1526 and gave his name to *Holbein Place* there. *Hunter Street*, Chester, bears the name of a cathedral verger who lived in the 18th century. The Liberal Member of Parliament William Hutt gave his name to *Hutt Street*, Hull (Hum), when it was laid out in 1850. *Hylton Road*, Petersfield (Ham), also built at this time, is named for another politician, William Jolliffe, first Baron Hylton, Conservative Member of Parliament for Petersfield. (His title came from Hylton Castle (T&W).)

Jack Straw's Lane, Headington, Oxford is popularly said to be named for Jack Straw, the priest who played a leading part in the Peasants' Revolt of 1381. He is thought to have stayed in a farmhouse here. Another account, however, takes the name from a wealthy farmer on Headington Hill. *James Street*, Lincoln, was originally *Vinegar Lane*, from a medieval form of the word vineyard. According to local archives, however, it was renamed in 1830 by a 'Captain Bromhead who adapted for the purpose his own Christian name, as he thought *Vinegar Lane* vulgar'. *Jenkins Lane*, Norwich, probably owes its name to a man who lived in this alley in the 17th century. More certainly, *Jowett Walk*, Oxford, was named in honour of Benjamin Jowett, Master of Balliol College. Its former name was *Love Lane* (see Chapter 3, p. 48). *Jordangate*, Macclesfield (Che), means 'Jordan's street', from one Jordan de Macclesfield, recorded as living here in the 14th century.

Kean Street, Covent Garden, London, commemorates the 19th-century actor Edmund Kean, who appeared at Drury Lane Theatre. *Keats Grove*, Hampstead, northwest London, is a reminder of the fact that the poet John Keats lived at Wentworth Place in this street for two years from 1818. The house is now the Keats Museum. *Knavesmire Road*, York, is a name meaning 'Knar's marsh', from the Old Norse personal name *Knorr* ('knot') and *myrr*, 'mire', 'marsh'. The fact that the region was a place of execution for criminals led to the alteration of the personal name to *Knave*. *Laburnham Road*, Maidenhead (Ber), is said to be named for L.A. Burnham, a builder of houses here. If so, it is a felicitous adoption, suggesting the popular flowering tree. *Ladbroke Grove*, Kensington, west London, is a reminder that Richard Ladbroke, a farmer, was a landowner here in the mid-18th century.

The rather cumbersome *Ladybellegate Street*, Gloucester, appears to be named for Lady Bell, wife of Alderman Sir Thomas Bell, here in the 16th century. More certain is the identity of the person who gave the name of *Lambs Conduit Street*, Holborn, north London. He was William Lamb, a musician by profession, who in the 16th century built a conduit here to bring fresh water from fields north of the City of London. The conduit was demolished in 1746. *Lansdowne Row* in Mayfair, west London, takes its name from William Petty, first Marquis of Lansdowne, better known by his Irish title of Lord Shelburne, who had a town house here in the 18th century. His English title derives from his estate at Lansdown, near Bath (Avn). *Lensfield Road*, Cambridge, takes its name from

John Lens, a fellow of Downing College for the first quarter of the 19th century, who owned a field nearby. *Lewin Street*, Middlewich (Che), has an association with an Anglo-Saxon named Lēofwine ('dear friend'), with the name recorded from the early 14th century. *Lillie Road*, Fulham, west London, is named for Sir John Lillie, who having served in the wars against Napoleon took an interest in creating this road in the mid-19th century. *Lordsmill Lane*, Chesterfield (Der), is more anonymously named for a mill that at one time belonged to the lord of the manor here.

Love Lane usually has the meaning discussed in Chapter 3, p. 48. In Petersfield (Ham), however, the street so named commemorates Robert Love, who owned land here in the early 18th century. *Lumby's Terrace*, Stamford (Lin), bears the name of Moses Lumby, a local butcher who acquired land here in the early 19th century. *Macaulay Road*, Clapham, southwest London, is not named for the famous historian but for his father, Zachary Macaulay, who joined the 'Clapham Sect' as a campaigner for the abolition of slavery. The somewhat exotically named *Margravine Road*, Fulham, west London preserves the title of the wife of the Margrave of Brandenburg-Anspach, who together with her husband bought a house here at the end of the 18th century. *Maud Heath's Causeway* is the name of a raised footpath running for some four miles by what is now the A420 road between Langley Burrell and Chippenham (Wil). It is said to take its name from a 15th-century marketwoman, who bequeathed her savings to build a paved causeway here from her home to Chippenham in order to avoid the muddy track that prevailed in her day.

Milsom Street, Bath (Avn), bears the name of a local wine cooper, Daniel Milsom, who rented land here in the mid-18th century. *Morrell Avenue*, Oxford, leading east to the suburb of New Headington, takes its name from the brewing family who lived at Headington Hill Hall. *Mosley Street*, Manchester, is named for the wealthy Mosley family, whose members included Sir Nicholas Mosley, Mayor of London in 1599, and Sir Oswald

Mosley, the 20th-century Fascist leader. *Moss Road*, Ilkeston (Der), bears the name of Alderman Horace Moss, Mayor of Ilkeston in the first quarter of the 20th century. *Much Lane*, Lincoln, preserves the name of the Much or Mutch family, here since at least the 17th century. *New George Street*, Hull (Hum), commemorates George Pryme, property owner and Member of Parliament for Cambridge in the 1830s. He also gave the name of *Pryme Street* here. 'New' distinguishes the former street from *George Street* (named for George III). In many towns *North Road* would indicate a geographical location or direction. In Berkhamsted (Her), however, it has the name of Joseph North, a local auctioneer and property developer.

Old Parr Road, Banbury (Oxf), rather unusually preserves the name of a murderer hanged here in 1747. *Opie Street*, Norwich, bears the name of the novelist Amelia Opie (1769-1853), widow of the painter John Opie, who had a house here. She may or may not have been aware that the medieval names of the street were *Gropecuntelane* (see Chapter 3, p. 49) and *Turpis Vicus*, 'dirty street', the Medieval Latin equivalent of the earlier vernacular. *Ottery Lane*, Sherborne (Dor), has the name of a family who originally came from Ottery (now Ottery St Mary) (Dev). *Paget Street*, Islington, north London, commemorates the distinguished surgeon Sir James Paget (1814-99), surgeon-extraordinary to Queen Victoria. *Peabody Avenue*, Westminster, west London, preserves the name of the American banker George Peabody, who in 1862 donated £150,000 to house the poor of London. His name was promoted by the many Peabody Buildings that were set up as a consequence. *Pearce Road*, Maidenhead (Ber), was named in memory of a 19th-century mayor, James Pearce. *Peck Lane*, Nottingham, bears the name of a medieval church official, John Peck. *Pegs Lane*, Hertford, is named for its builder, William Pegg. The name behind *Picklers Hill*, Abingdon (Oxf), is actually *Pulcher*, that of a local family.

Pimlico Walk, Hoxton, north London, is a reminder that Ben Pimlico was an innkeeper here in the 16th century. His beer was popular, and was sold in other inns to the west of the City in what is now the district of Pimlico. *Plumptre Street*, Nottingham, is not so called for its plum trees but for a local family named Plumptre. *Pope Street*, Bermondsey, south London, takes its name from Sir Thomas Pope, who held the manor of Bermondsey in the 16th century and who built Bermondsey House here with the stones of the old Bermondsey Abbey. *Praed Street*, Paddington, west London, was built on land bought in the early 19th century by the Grand Junction Canal Company, whose chairman was William Praed. *Pullen's Lane*, Oxford, running between the top of Headington Hill and Jack Straw's Lane, takes its name from Josiah Pullen (1631-1714), a local vicar and vice-president of Magdalen Hall, an academic hall on the site of the present Magdalen College.

Rathbone Place, off Oxford Street, west London, was built by Thomas Rathbone in the early 18th century. *Rolleston Street*, Salisbury (Wil), is said to take its name from one Rolfe, who built houses here in medieval times, so that it was 'Rolfe's town'. The same man may have given the name of *Rollestone*, 10 miles northwest of Salisbury. *Ronalds Road*, Islington, north London, commemorates the scientist and inventor Sir Francis Ronalds (1788-1873), who lived in nearby Highbury Terrace. *Rosoman Street*, a short distance to the south, preserves the name of Thomas Rosoman, first proprietor of Sadler's Wells Theatre and builder of the street.

St John's Hill, Wandsworth, southwest London, is not named for a local church but for the St John family, who held the manor of Battersea here in the 17th and 18th century. *Scots Lane*, Salisbury (Wil), is said to take its name from the family of Antony and William Scot, recorded here in medieval times. *Serjeant Street*, Peterborough (Cam), bears the name of Joseph Serjeant, a local wholesale butcher. London's famous *Shaftesbury Avenue* was built in the 1880s to link New Oxford Street with Piccadilly and was named for the seventh Earl of Shaftesbury, whose charitable work had largely been directed towards helping the poor in this former slum area. *Shepherd Market*, off

Piccadilly, west London, was built by Edward Shepherd in the early 18th century. *Sheraton Street*, Soho, west London, commemorates the famous cabinetmaker Thomas Sheraton, who lived here and in nearby Wardour Street in the closing years of the 18th century. The 18th-century essayist Richard Steele is remembered in the name of *Steele's Road*, Chalk Farm, northwest London. The writer had a cottage here at the junction with Haverstock Hill. *Stepping Lane*, Norwich, first recorded in the 18th century, appears to derive from the name of a local family. The East Anglian surname is more familiar as *Stebbing*. *Storer Road*, Loughborough (Lei), preserves the name of John Storer, a local benefactor who died in 1713. *Stukeley Street*, Camden, north London, serves as a memorial to the eccentric antiquarian William Stukeley (1687-1765), who lived in Holborn.

Timbrell Street, Trowbridge (Wil), bears the name of Thomas Timbrell, to whom the manor of Trowbridge passed in 1807. *Train Avenue*, Hull (Hum) is nowhere near the railway but has preserved the name of Norman Train, a building manager on this estate street. *Tulse Hill*, Lambeth, south London, is a reminder of the Tulse family, owners of Tulse Hill Farm here in the 17th century.

Walmgate, York, recorded in the 11th century as *Walbegate*, appears to derive its name from one *Walba*, a short form of the Anglo-Saxon name *Wealhbeorht*. The *b* must have become *m* for ease of pronunciation. *Walpole Street*, Chelsea, west London, commemorates Sir Robert Walpole, England's first Prime Minister, who is said to have lodged here before leasing Walpole House in 1724. *Walwers Lane*, Lewes (ESx), preserves the name of William le Walewere, here in the 13th century. *The Wardwick*, Derby, recorded in the late 11th century as *Walwik streete*, has a name that means 'Walda's farm'. The *l* of the name became *r* some time in or before the 18th century. The hymnwriter Charles Wesley is buried in the churchyard of the now demolished Marylebone parish church in Marylebone High Street, and is remembered in *Wesley Street* here in northwest

London. Finally in this extensive selection, *Zinzan Street*, Reading (Ber), preserves the name of a Mrs Zinzan, recorded as living here in 1800.

12. COMMEMORATIVE AND PROPITIOUS NAMES

Most of the names cited in the previous chapter are personally and locally commemorative, since they almost entirely relate to people, some now little known, others still famous, who were individually connected with the street.

A true commemorative name, however, may be said to be one that is given either to mark a national occasion, such as a royal birth or a victory in a battle, or generally as a compliment to a currently important person, such as a reigning monarch, an officiating prime minister, or even a foreign head of state or 'popular hero'. The familiar *Coronation Street*, found in many towns and cities, is just one example of a name that commemorates a unique royal event, and the name is usually given to a street that is laid out or built at the time. Commemorative names thus involve equally both people and events. In modern times the naming of a street for a twin town in a foreign country can be regarded as a special kind of commemorative name, since it honours the other town and keeps its name actively alive in the local public consciousness (so long, of course, as the public is aware of the significance of the name!) Looked at another way, a true commemorative name usually has no actual physical link with the particular street at all. In some instances, however, the royal or other personage in question may have paid an official visit to the town, or even opened the street, as happened with *Clarence Street*, Kingston upon Thames, for example (see this name below).

Royal commemorations are one of the most readily identifiable types of street name. (Bridges with names of this kind were mentioned in Chapter 10.) The name may consist of a royal personal name or else a title. *Adelaide Street*, off the Strand, west London, was built in 1831 and named for Queen Adelaide, wife of William IV. *Albany Street*, Camden, north London, was built down the eastern side of Regent's Park at the same time as *Regent Street* (see below) and was named for the Prince Regent's brother, Frederick, Duke of York and Albany. (The various Terraces and Gates surrounding the Park bear relevant royal titles similarly, running anticlockwise alphabetically from the southeast corner: *Cambridge Terrace*, *Chester Terrace*, *Cumberland Terrace*, *Gloucester Gate*, *Hanover Terrace*, *York Terrace*. *Cornwall Terrace* is at the southwest corner. The Prince Regent was not only Prince of Wales but Duke of Cornwall and Earl of Chester.) London's *Albert Embankment*, on the south bank of the Thames, was built in the 1860s and named in memory of the recently deceased Prince Albert (died 1861), husband of Queen Victoria. Facing it across the river is the *Victoria Embankment*, built at the same time (see also *Victoria Street*, below). *Alexandra Terrace*, Lincoln, was so named in 1874 for Princess Alexandra, wife of the Prince of Wales (she was later Queen Alexandra while he was Edward VII). *Brunswick Square*, to the west of Coram's Fields, north London, was laid out in the late 18th century and named as a compliment to George III of the house of Hanover, for whom Brunswick was the second capital. *Mecklenburgh Square*, east of the Fields, was laid out at the same time and named for the king's wife, Queen Charlotte, formerly Princess Charlotte of Mecklenburg-Strelitz.

A more original royal commemoration is that of *Canute Place*, Knutsford (Che). Until the 19th century this was the unremarkable *Market Place*. The local authorities were then persuaded that the name of the town, recorded in the Domesday Book of 1086 as *Cunetesford*, derived from that of King Canute (died 1035), who must therefore be connected with it in some way. What more logical and fitting than to rename the town's principal square for him? Although the town's name does mean 'Knut's ford', and this Old Danish name (meaning 'knot') is the same as that of the king, there is no evidence that it was he who gave the town's name. (He is said to have died at Shaftesbury (Dor) and is buried at Winchester (Ham).)

Charlotte Street, Camden, north London, was laid out in the late 18th century and named for Queen Charlotte, wife of George III. *Clarence Street*, Kingston upon Thames, southwest

London, is named for the Duchess of Clarence, later Queen Adelaide (see *Adelaide Street*, above), who actually opened it in 1828. *Cobourg Street*, Euston, north London, commemorates the marriage of Princess Charlotte (not the above), daughter of George IV, to Prince Leopold of Saxe-Coburg in 1816. The death of the popular princess in childbirth the following year was regarded as a national tragedy. (The street has a French spelling of the name of the German city that gave the name of the duchy.) *Cumberland Gate*, by Marble Arch, west London, was originally *Tyburn Gate*, named for the small stream here. It was renamed in honour of Prince William, brother of George III and victor at the Battle of Culloden in 1746. *Denmark Street*, off Charing Cross Road, north London, was named for Prince George of Denmark, husband of Queen Anne, who died in 1708. He is said to have also given the name of *Denmark Hill*, Camberwell, southeast London, from his fondness for hunting there. *Duke of York Street*, Westminster, west London, bears the name of James, Duke of York, brother of Charles II. The street appropriately leads to *St James's Square*, although that was named for St James's Palace, itself named for a leper hospital. *Ebor Street*, Bethnal Green, east London, was originally *York Street* for the same duke. As his name abounded in those of London streets, however, it was given a variant name, derived from the abbreviation for *Eboracum*, the Roman name of York.

Elizabeth Avenue, Abingdon (Oxf), is just one of many streets so named in 1977, the year of the silver jubilee (25th anniversary of the accession) of Queen Elizabeth II. *Jubilee Street*, Peterborough (Cam), however, as frequently elsewhere, was named for the golden jubilee (50th anniversary of the accession) of Queen Victoria, in 1887. However, *Jubilee Street*, Stepney, east London, was named to commemorate the golden jubilee of George III in 1810. This same king gave the name of *Great George Street*, Weymouth (Dor), as a result of his official visit to Melcombe Regis, now a district of Weymouth, in 1789. *Royal Crescent*, also in Weymouth, is another name commemorating the king's visit. *Hanover Square*, off Oxford Street, west

London, was laid out in 1713 by the Earl of Scarborough, a keen supporter of the house of Hanover and the Protestant succession. The name directly compliments George I, as does *St George Street* (earlier *George Street*), which leads to it. *Henrietta Street*, Covent Garden, west London, was laid out in 1631 and named for Henrietta Maria, wife of Charles I. *Kent Terrace*, to the west of Regent's Park, is named for the brother of the Prince Regent, Edward, Duke of Kent. (See also *Albany Street*, above.)

King Ecgbert Road in Dore, now a southwest district of Sheffield (SYo), commemorates the treaty between this king of the West Saxons and the Northumbrians here in 829. *King Edward Street*, in the City of London, is named for Edward VI, on the throne when the Greyfriars monastery here was converted into Christ's Hospital as a home for poor children. *King Square*, Islington, north London, no longer a square as such, was so named in 1820 for George IV, who succeeded to the throne that year on the death of his father, George III. *King's Road*, Chelsea, west London, fashionably famous in modern times, is recorded in 1620 as *The King's High Way from Chelsea to London* and in 1717 as *The King's Private Road*. It was thus the royal route from St James's Palace and what was then Buckingham House (and is now Buckingham Palace) to Hampton Court, crossing the river by ferry at the point where Putney Bridge now stands. The road became a public highway only in 1830. *Kingsway*, leading from Southampton Row to Aldwych, west central London, was built over the period 1899 to 1906 and named for Edward VII, reigning from 1901 to 1910.

Queen Anne's Gate, Westminster, west London, approaches the gate into St James's Park that was named for Queen Anne. The former name of the street was *Queen Square*. *Queen Anne Street*, to the north of Oxford Street, west London, is named for the same monarch. The street was built to lead to a square that in the event never materialised. *Queen Elizabeth's Walk*, Stoke Newington, north London, is more relevantly named for Queen Elizabeth I, who in the 16th century visited John Dudley, Duke of Northumberland and Earl of Warwick, at Stoke Newington

Manor House here. Many streets with this name, with or without its royal prefix, pay tribute to Queen Elizabeth II, however (like *Elizabeth Avenue*, above). *Queen Square*, Holborn, north London, was laid out in the early 18th century and named for Queen Anne. *Queen Square*, Bath (Avn), on the other hand, was completed in 1734 and named for Queen Caroline, wife of George II. *Queen Street*, Oxford, again, was named for Queen Charlotte (see *Charlotte Street*, above), who visited Oxford with her husband in 1785. *Queensway*, Westminster,- west London, was named in the 19th century (originally as *Queen's Road*) for Queen Victoria, as was the identically named street in Hemel Hempstead (Her), and doubtless in many other towns. London's *Queen Victoria Street*, built from the Bank of England to Blackfriars Bridge in 1871, is a royal name that is self-explanatory.

Regent Street, Westminster, west London, was built over the period 1813 to 1821 as part of a scheme for the Prince Regent (the future George IV). The street was intended as an approach to Regent's Park (see *Albany Street*, above), with *Lower Regent Street* leading to Carlton House, the Prince's residence. Insofar as the scheme was part of that designed by the architect John Nash to connect the former Marylebone Park (now Regent's Park) with the Prince Regent's house, the name was personally relevant, rather than generally commemorative. At the same time the new street would have been planned and built anyway to link the Park with central London, and its name was thus intended to preserve the title of the (effective) monarch then reigning.

Streets named *Royal* are often named for a building or an institution (such as London's Royal Hospital) rather than for a monarch or member of the royal family. However, *Royal Arcade*, the shopping arcade off Old Bond Street, west London, is said to owe its name to a visit here from Queen Victoria, and the famous *Royal Crescent* at Bath (Avn) has a name that relates as much to its grandeur as to the city's royal historic connections.

Victoria Street, Westminster, west London, was built in 1851 and named for Queen Victoria, who thanks to her personal prestige and long reign has arguably given her name to more places, from the largest of states to the shortest of streets, than any other British monarch. Greater London currently has over 20 streets named for her, mainly as *Victoria Road*. It was this particular *Victoria Street* that gave the name of Victoria Station, and the station that in turn passed its name to the district here.

Coronation Street (or *Coronation Road*), already mentioned, is found in many towns. It usually marks one of the four coronations that have occurred in Britain since the beginning of the 20th century, viz of Edward VII (1902), George V (1911), George VI (1937) or Elizabeth II (1953). (Edward VIII was never crowned.) *Coronation Street*, Ilkeston (Der), for example, was so named for the crowning of Edward VII, and *Coronation Road*, Hull (Hum), for that of George VI.

Apart from royalty, streets have been named commemoratively for a whole range of famous people. Some of them, such as musicians and actors, appear in the names cited in the previous chapter. Elsewhere, among the expected military heroes (*Gordon Road*, Peckham, southeast London, *Marlborough Place*, St John's Wood, northwest London, *Nelson Square*, Southwark, south London, *Wellington Road*, St John's Wood, northwest London) and battle victories, one finds such diverse commemorations as *Anna Neagle Close*, Forest Gate, northeast London, *Nell Gwynne Avenue*, Shepperton (Sur), *Nelson Mandela Close*, Muswell Hill, north London, and *Winston Churchill Avenue*, Portsmouth (Ham). (See Chapter 13, p. 179, for further names of this type.) One of the most self-evident commemorative names in the country must be that of the *Road of Remembrance*, Folkestone (Ken). The road, planted with rosemary (for remembrance), commemorates the fallen of the First World War.

Battle victories, just mentioned, have provided a host of commemorative street names, and there is hardly a battle or siege that will not have a street somewhere named for it. Battles and sieges of the Peninsular War, Crimean War and Boer War are particularly well represented, with First and Second World War victories also found. It can normally be taken that the street was built or renamed shortly after the date of the battle. The following (with date of battle) are found by way of example in Greater London and the Home Counties: *Alamein Gardens* (1942), Dartford, *Albuhera Close* (1812), Enfield, *Aliwal Road* (1846), Battersea, *Alma Grove* (1854), Southwark, *Armada Street* (1588, Deptford, where many ships were fitted out to fight the Spanish fleet), *Arnhem Drive* (1944), Croydon, *Balaclava Road* (1854), Southwark, *Belgrade Road* (1789), Hackney, *Blenheim Road* (1704), Epsom, *Corunna Road* (1809), Battersea, *Inkerman Road* (1854), Woking, *Ladysmith Avenue* (1900), Ilford, *Mafeking Avenue* (1900), Brentford, *Maida Vale* (1806), Westminster, *Marne Avenue* (1918), Friern Barnet, *Plevna Road* (1877), Hampton, *Pretoria Avenue* (1900), Walthamstow, *Ramillies Place* (1706), Soho. Of these, *Blenheim* easily leads the field in ubiquity.

The most famous of London's battle street names, however, is that of *Trafalgar Square*, laid out in 1830, but commemorating Nelson's victory a quarter of a century earlier. The name is not of course unique to London, and is represented elsewhere by *Trafalgar Road*, Gravesend (Ken), for example, or *Trafalgar Street*, Burnley (Lan).

A modern type of commemorative name is that of a foreign twin town. Many English towns maintain cultural links with a European town, especially one in France or Germany, and name a street to mark the particular link and preserve the name of the 'twin'. Mention has already been made of *Alencon Link*, Basingstoke (Ham), *Bourges Boulevard*, Peterborough (Cam), and the French-style *Avenue de Chartres*, Chichester (WSx). Further examples are *Angouleme Way*, Bury (GtM), *Arles Avenue*, Wisbech (Cam), *Armentieres Square*, Stalybridge (GtM), *Bonn Square*, Oxford, *Chalon Way*, St Helens (Mer) (for Chalon-sur-Saône), *Dinant Link Road*, Hoddesdon (Her), *Epinal Way*, Loughborough (Lei), *Evry Road*, Bexley (GtL),

Freetown Way, Hull (Hum) (for the capital of Sierra Leone), *Herford Way*, Hinckley (Lei), *Kranj Way*, Oldham (GtM), *Lucerne Avenue*, Bournemouth (Dor), *Marne Close*, Ilkeston (Der) (for Châlons-sur-Marne, although suggesting a battle name), *Minden Road*, Sutton (GtL), *Mons Square*, Bootle (Mer), *Paderborn Court*, Bolton (GtM), *Pontoise Close*, Sevenoaks (Ken), *Rennes Drive*, Exeter (Dev), *Rheims Way*, Canterbury (Ken), *Roth Drive*, Brentwood (Ess), *Rouen Road*, Norwich, *Stains Close*, Cheshunt (Her) and *Vence Walk*, Stamford (Lin). Despite appearances, *Noreuil Road*, Petersfield (Ham), is not named for a twin town but for the village near Arras, northeast France, whose reconstruction after the First World War was aided by Petersfield. *Noreuil Road* itself developed from the need to provide housing for ex-servicemen at this time.

Sometimes, especially when the street is of some importance, it will be ceremonially opened and named jointly by officials from the English town and its 'twin'. A plaque by Oxford's *Bonn Square* thus tells how: 'The Lord Mayor of the City of Oxford, Councillor Mrs Olive Gibbs, and the Chairman of the Bonn City District Council, Herr Stadtverordneter Reiner Schreiber, named this Square on 5 October 1974'.

Many overseas towns have reciprocated the compliment, and have named their streets for their English 'twins'. Leningrad (now St Petersburg), for example, twinned with Manchester in 1956 and so has a *Manchesterskaya ulitsa* ('Manchester Street'), although Manchester itself now has no street named for its Russian counterpart.

It is not unusual for an English town to be twinned with more than one overseas town (Coventry (WMd) is linked with no less than 20), and they may all be represented in street names, typically on a new housing estate. Abingdon (Oxf) is twinned with six towns, so duly has *Argentan Close*, *Bergen Avenue* (for Bergen-op-Zoom), *Coromandel* (*sic*, lacking generic term, for the New Zealand district where Abingdon's twin town of Thames is situated), *Lucca Drive*, *Schongau Close* and *Sint Niklaas Close*, while Maidenhead (Ber) boasts its three twinnings in *Bad*

Godesberg Way, *Frascati Way* and *St Cloud Way*. Abingdon also has its *Virginia Way*, for its American namesake, the town of Abingdon in Virginia. The two communities are not officially twinned, however.

The earliest such names date from the 1940s, when the twin-town movement was inaugurated in a spirit of postwar reconciliation. Accents are usually ignored in the English spellings (eg for Alençon, Angoulême and Armentières), and the pronunciation of the names is also normally anglicised.

In a few instances a town or district has commemorated itself, so to speak, by using its own historic name as a street name. Thus tiny *Battle Bridge Road*, to the west of King's Cross Station, London, preserves the name of the former district here. (*Battle Bridge* is a corruption of *Bradford Bridge*, referring to the bridge that superseded the original 'broad ford' here over the Holborn.) *Cheniston Gardens*, Kensington, south London, has a name based on *Chenesitun*, the name of Kensington as recorded in the Domesday Book, while *Ellington Park*, Maidenhead (Ber), preserves the former name of Maidenhead, recorded in the Domesday Book as *Elentone. Glevum Close*, Gloucester, revives the city's Roman name, as does *Lindum Road*, Lincoln. *Thorney Street*, Westminster, west London, commemorates the former island of Thorney ('thorn island'), a portion of firm ground between the Thames and Tyburn Brook where an early monastic settlement was established on a site that is now that of Westminster Abbey. As in Lincoln, so in St Albans (Her) *Verulam Road* preserves the Roman name of the city, *Verulamium*. On similar lines *Lolesworth Street*, Spitalfields, east London, is a reminder of the former name of Spitalfields, recorded in 1278 as *Lollesworthe*, a field name (like that of Spitalfields itself), while *Paddenswick Road*, running down the eastern side of Ravenscourt Park, Hammersmith, west London, commemorates the old name of the district here, recorded variously as *Palyngeswyk* in 1270, *Paddyngeswyke* in 1553, and *Padderswick* in 1819. *Tyburn*, the former name of Marylebone, northwest London, is likewise preserved in *Tyburn Way*. The street is not in Marylebone, however, but at Marble Arch, west London, on the site of the former notorious Tyburn gallows. Even so, the district was named for the small river *Tyburn*, which flowed through what is now Marylebone and under today's Oxford Street, so that the siting of the modern street name is relevant both geographically and historically. (It was the renaming of the stream as the *Marybourne* that produced the new district name.)

A royal street name is a prestigious one, even if (or because) the street itself is not particularly remarkable. But there are several streets whose names are 'classy' in some other way, often arbitrarily. *The Paragon*, Bath (Avn), a street built in 1769, is one example of such a name, since the word suggests something that is a model of its kind. Other names are more propitious than prestigious. Any name suggesting a rural location, for example, is of this kind, whether the actual locality is truly rural or not. *Springfield Road* is a common example, although it is unlikely that all streets of this name were originally laid out by a spring in a field or by a field called *Springfield.* Similarly any name suggesting a pleasant place is often given simply for its associations. For instances of rural names of this kind, see Chapter 13, p. 179.

The best-known example of a propitious name is probably *Mount Pleasant*, suggesting an elevated and salubrious locality that is agreeable to live in. Yet this very name was often given ironically, to a place that was noticeably unsavoury or unhealthy. One famous instance is that of *Mount Pleasant*, Clerkenwell, north London, linking Gray's Inn Road with Farringdon Road, familiar from the Mount Pleasant Sorting Office here. The name is first recorded in 1732 for a lane that led to a huge mound of refuse or cinders at the Gray's Inn Road end. Other streets of the name with similar mounds existed elsewhere in London, and the mounds themselves are mentioned in Dickens' *Our Mutual Friend* (1864) as making the fortune of Nicodemus Boffin, the 'Golden Dustman'. *Laystall Street*, with its more factual name (see Chapter 5), led to the Clerkenwell refuse dump just mentioned.

Another name of this type is *Paradise Street*. The name could be ironic, but sometimes originated from that of a field or a house, or related to a garden or, in a religious sense, to a burial ground. *Paradise Square*, Oxford, was mentioned in Chapter 4 as deriving from the name of an orchard or 'garden of paradise'. On the other hand, *Paradise Walk* in Chelsea, west London, originally *Paradise Row*, was so named for its small but elegant Georgian houses. It was simply a pleasant street to live in.

At the time of the Industrial Revolution, in the 18th and 19th centuries, many expanding and flourishing towns adopted the names of well-known London streets, seeing themselves as increasingly important commercial rivals to the capital. With industrial competition, too, went cultural ambition, so that the names adopted were the prestigious ones. Such names are particularly in evidence in northern cities, where industrial expansion was most marked. Manchester, for example, has its *Chancery Lane*, *Fetter Lane*, *Fleet Street*, *Great Marlborough Street*, *King Street*, *Oxford Street*, *Pall Mall*, *Piccadilly*, *Portland Street* and *Spring Gardens*, Liverpool has *Bond Street*, *Cheapside*, *Covent Garden*, *Drury Lane*, *Fleet Street*, *Hatton Garden*, *Islington*, *Oxford Street*, *Pall Mall*, *Park Lane*, *Parliament Street*, *The Strand*, *Soho Street*, *Vauxhall Road*, *Wapping* and *Whitechapel*, while further south, in the Midlands, Birmingham streets include *Cannon Street*, *Cheapside*, *Fleet Street*, *Sloane Street* and *Snow Hill*. This last name, thanks to the well-known former railway station, became generally even more familiar than its London equivalent, a shortish street in the City linking Farringdon Road and Holborn Viaduct.

It should be said, however, that such borrowings were often relevant to local circumstances. Manchester's *Chancery Lane*, for instance, while acknowledging a debt to London, commemorates the fact that the estates of a local landowner, one Robert Hamilton, were dealt with in the Court of Chancery. *Spring Gardens*, in the same city, was actually laid out next to a well-known local fountain, while *Oxford Street* was one of a number of 'superior' names selected almost at random for a group of streets round *Great Marlborough Street*. Others here included *Cambridge Street*, *Cavendish Street*, *Grosvenor Street*, *Ormond Street* and *York Street*. Ironically, these aristocratically named streets came to be the centre of one of Manchester's most notorious slum districts, singled out for denunciation by Friedrich Engels in *The Condition of the Working Class in England in 1844* (1845). Again, even Manchester's *Piccadilly*, so named in 1812, could claim a local association with its London namesake in that both streets had a well-known coaching inn and that the inn in each case was the White Bear. The 'London link', however, would have been an important factor in these names and others like them found elsewhere.

13. THEMATIC NAMES

Open almost any street plan of any town or city, and you cannot help noticing groups of related names in one or other part of the town, often in a run of Victorian terraced streets or on a modern housing estate. When districts like this are laid out, the bestowal of associated names is an obvious solution to the problem of individually identifying each of the new streets while at the same time having a uniting theme that will point to their contemporaneity. Thus Staines (Sur) has its arboreal clump of *Ash Grove*, *Beechwood Avenue*, *Cherry Tree Avenue*, *Chestnut Grove*, *Hazel Grove* and *Hurstdene Avenue*, while Bootle (Mer) has its Oxford University college group of *Balliol Road*, *Exeter Road*, *Hertford Road*, *Keble Road*, *Oriel Road*, *Pembroke Road*, *Queen's Road*, *Trinity Road* and *Wadham Road*. The latter are complemented to their east by a Cambridge college cluster of *Clare Road*, *Downing Road*, *Gonville Road* and *Sidney Road*, either side of *Cambridge Road*.

In some cases, thematic names of this type are little more than randomly appropriate. Rural names, for flowers and trees, are extremely popular, not only because they are pleasantly evocative but because many new housing estates are laid out at the edge of a town as a sort of 'garden city' incursion into the surrounding countryside. (In the tree names above, even the generic names *Avenue* and *Grove* are apt.) In other instances the names will have a local significance, perhaps one not immediately apparent to the outsider.

Some thematic groupings of local reference can be fairly complex, and involve a combination of associated personal names and place-names. In Marylebone, northwest London, there are many streets named for the family and descendants of John Holles (1662-1711), Duke of Newcastle, who gained his title after his marriage in 1689 to Lady Margaret Cavendish, daughter and heiress of Henry Cavendish, second Duke of Newcastle, himself related to the Earl of Devonshire. In 1713 Holles' own daughter, Henrietta, married Edward Harley (1689-1741), second Earl of Oxford and Mortimer, otherwise

Baron Harley of Wigmore Castle, who had property at Wimpole (Cam), north of Royston. In 1734 Harley's daughter Margaret in turn married William Bentinck (1708-62), second Duke of Portland, whose seat was at Welbeck, near Worksop (Not), and who had property at Carburton and Clipstone, respectively east and south of Welbeck. Their own eldest daughter married Thomas Thynne, second Viscount Weymouth. The names of the family members and of their properties gave those of the following streets, many of them well known in London's West End: *Bentinck Street*, *Carburton Street*, *Cavendish Square*, *Clipstone Street*, *Devonshire Street*, *Great Portland Street*, *Harley Street*, *Henrietta Place*, *Holles Street*, *Margaret Street*, *Mortimer Street*, *Oxford Street* (but see also Chapter 5, p. 62), *Portland Place*, *Welbeck Street*, *Weymouth Street*, *Wigmore Street* and *Wimpole Street.* If Holles had become Duke of Clarence, as was his wish, there would undoubtedly have been a 'Clarence Street' here as well. The names are all located in this particular part of London because Holles had bought lands in Marylebone in 1708.

Similar 'linkage' is found in several other towns. In Ilkeston (Der), for example, many names relate to the local Manners family, as Dukes of Rutland and Marquises of Granby. Prime references are preserved in the names of *Duke Street*, *Granby Street*, *Manners Avenue* and *Rutland Street. Belvoir Street* is named for their seat at Belvoir Castle (Lei). *Vernon Street* commemorates the Vernon family of Haddon Hall, near Bakewell (Der). Dorothy Vernon, daughter of Sir George Vernon (1578-1639), married into the Manners family, who in consequence inherited Haddon Hall. Their property there is preserved in the name of *Haddon Street. Bristol Road*, *King Street*, *Scarborough Avenue* and *Wharncliffe Road* bear the names of agents and trustees appointed by the Manners family, with two of the trustees being the Earl of Scarbrough (*sic*) and Earl of Wharncliffe. *Peacock Place* is said to take its name from the peacock which forms part of the Manners family crest. The nearby Peacock Colliery was owned by one of the Dukes of Rutland, and the Peacock Inn in Ilkeston is named for them.

Sometimes the grouping may consist simply of personal names. Abingdon (Oxf), for instance, has neighbouring streets named for former mayors: *Allder Close*, *Eldridge Close*, *Franklyn Close*, *Kysbie Close*, *Langley Road* and *Spenlove Close*. The same town, however, has a quite different grouping of personal and place-names that exploits local associations with the motor industry, and particularly with the MG sports car manufactured in Abingdon from 1930 to 1980. It includes *Eyston Way*, named for Captain Eyston, racing car driver for the MG manufacturers in the 1930s, *Healey Close*, for the Austin Healey car made at the MG works in the 1950s, *Kimber Road*, for Cecil Kimber, who set up the MG factory in the town, *Midget Close*, for the MG 'TF' sports car, whose small wheelbase earned it the nickname 'Magic Midget', *Nuffield Way*, for William Morris, Lord Nuffield, founder of the MG (ie Morris Garages) Car Company in 1930, and *Riley Close*, for the Riley tourer car, made by the Riley Company, Coventry, which Lord Nuffield acquired in 1938.

Another grouping of combined personal and place-names, although quite different in character, is found in Borehamwood (Her), with its Elstree film studios. Here the streets adjacent to *Studio Way* display a generous cinematic representation: *Banks Road*, *Buchanan Court*, *Cygnet Close*, *Danziger Way*, *Denham Way*, *Ealing Close*, *Hancock Court*, *Hawkins Close*, *Hay Close*, *Lion Court*, *Mason Close*, *Neagle Close*, *Pinewood Close*, *Shepperton Close* and *Wilcox Close*. In deploying the names of film actors, producers and studios on this estate, the local planners took pains to site *Neagle Close* opposite *Wilcox Close*, so that the film producer and director Herbert Wilcox should be facing his wife, the actress Anna Neagle.

Victorian north Oxford has an impressive region of streets named for properties or livings owned by St John's College. As well as *Bainton Road*, mentioned in Chapter 11, p. 146, they include the following: *Bardwell Road*, *Belbroughton Road*,

Bevington Road, *Bradmore Road*, *Chadlington Road*, *Chalfont Road*, *Charlbury Road*, *Crick Road*, *Farndon Road*, *Fyfield Road*, *Garford Road*, *Kingston Road*, *Leckford Road*, *Linton Road*, *Northmoor Road*, *Polstead Road* and *Warnborough Road*. Some of these place-names are far from local. *Belbroughton Road*, for example, derives from a living in the village of this name near Stourbridge (H&W), while *Bardwell Road* is named for a village near Bury St Edmunds (Suf). *Bevington Road* was earlier *Horse and Jockey Road*, after a pub here. In 1968 St Anne's College, owner of houses in this road, asked the Highways Committee of the City Council whether its name could be changed to *Plumer Road*, for a former principal of the college, the Hon. Eleanor Mary Plumer, who had died the previous year. Permission was not forthcoming, however.

The motor industry names quoted for Abingdon above are not grouped together in one part of the town but are fairly widely distributed. A similar widespread diffusion of such names can occur in quite a large town where there is a well established local reputation or association. One example is the well-known military town of Aldershot (Ham), where there are many names representing battle victories and war leaders. Among them are *Alamein Road*, *Anzio Close*, *Arnhem Close*, *Cassino Close*, *Churchill Avenue*, *Crimea Road*, *Falaise Close*, *Gordon Road*, *Imjin Close*, *Kohima Close*, *Redan Hill*, *Roberts Road*, *Salerno Crescent*, *Sebastopol Road*, *Waterloo Road*, *Wellington Avenue* and *Wolfe Road*, as well as *Gun Hill* and *Ordnance Road*.

A less common thematic device is to vary the generic term, not the specific. One example was the grouping of *Castle* names cited for Kendal (Cum) in Chapter 1, p. 2. Another, if anything more spectacular, is found in Leeds (WYo), where the Middleton Park district has *Throstle Avenue*, *Throstle Drive*, *Throstle Grove*, *Throstle Hill*, *Throstle Lane*, *Throstle Mount*, *Throstle Parade*, *Throstle Place*, *Throstle Road*, *Throstle Row*, *Throstle Square*, *Throstle Street*, *Throstle Terrace*, *Throstle View* and *Throstle Walk*, a total of 15 different terms.

THEMATIC NAMES

The following are some of the general themes most commonly found in English street name groups, with examples for each. There may or may not also be a local reference, and the names quoted may not always be completely comprehensive for the group.

aircraft: *Anson Way, Blenheim Drive, Canberra Road, Comet Way, Lancaster Drive, Lincoln Way, Lysander Close, Stirling Road, Valiant Road, Victor Close, Wellington Avenue,* St Ives (Cam) (near airfields).

alphabetical identity: *Oakland Avenue, Oakville Drive, Odessa Avenue, Orama Avenue, Orient Road, Orlanda Avenue, Orme Avenue, Ormonde Avenue, Orvietto Avenue, Otranto Avenue,* Eccles (GtM) (one of these, perhaps *Oxford Road,* may have engendered the others).

alphabetical order: *Avenue Road, Blythe Road, Clyde Road, Derby Road, Eldon Road,* Nazeing (Ess).

American states and cities: *Albany Place, Boston Street, Carolina Street, Concord Place, Florida Street, Jersey Street, Nevada Street, Oregon Street,* Bootle (Mer).

ancient history and mythology: *Apollo Close, Augustan Close, Barrow Close, Britannia Way, Caesar Green, Capitol Close, Celtic Crescent, Claudius Close, Diana Close, Druids Walk, Forum Green, Fosse Green, Gladiator Green, Legion Close, Martyr Close, Minerva Close, Mithras Close, Remus Close, Roman Road, Romulus Close, Temple Close, Vespasian Way,* Dorchester (Dor) (a Roman town).

artists: *Constable Road, Da Vinci Close, Degas Drive, Fraser Drive, Gainsborough Drive, Hogarth Close, Holbein Road, Lowry Close, Monet Close, Morland Way, Rembrandt Way, Renoir Close, Reynolds Close, Romney Close, Rubens Way, Spencer Drive, Stubbs Close, Turner Road, Vandyke Place, Van Gogh Place, The Whistlers,* St Ives (Cam) (although with such names one would almost have expected St Ives (Cor), a town famous for its artists).

astronomy: *Altair Way, Apollo Avenue, Aquarius Way, Aquila Close, Avior Drive, Canopus Way, Capella Road, Orion Way, Phoenix Close, Vega Crescent,* Northwood, northwest London.

battles: see Aldershot (Ham) names, above, p. 182.

birds: *Falcon Way, Fulmar Head, Gold Crest, Great Auk, Heron Gate, Kestrel Hide, Kingfisher Drive, Lark Drive, Little Grebe, Mallard Close, Merlin Close, Montagu's Harrier, Osprey Close, Raven Close, Redwing Rising, Swallow Close,* Guisborough (Cle) (with some of the bird names serving unusually as generic street names: *Grebe* was presumably designed to echo *Glebe*).

British Empire or Commonwealth: *Australia Road, Canada Way, Commonwealth Avenue, India Way, New Zealand Way, South Africa Road,* Hammersmith, west London (near White City, where Olympic Games were held).

cathedral cities: *Canterbury Road, Chester Close, Coventry Road, Edinburgh Drive, Ely Close, Gloucester Close, Guildford Close, Hereford Road, Lichfield Road, Lincoln Road, Norwich Road, Peterborough Road, Truro Drive, Winchester Avenue,* Exeter (Dev) (a cathedral city).

counties: *Cornwall Gardens, Cumberland Avenue, Devon Road, Dorset Road, Essex Road, Hampshire Road, Kent Avenue, Rutland Close, Somerset Road, Suffolk Road, Surrey Road, Sussex Avenue, Warwick Road,* Canterbury (Ken) (compare **dukes**, below; the names often overlap).

country houses and castles: *Arundel Close, Charlecote Avenue, Chatsworth Avenue, Denham Close, Glencairn Avenue, Harewood Close, Longleat Avenue, Osborne Avenue, Petworth Close, Sandringham Avenue, Sulgrave Close, Thoresby Avenue, Warwick Avenue, Windsor Drive, Woburn Avenue,* Gloucester.

directional: *Aberdeen Close, Angus Close, Argyll Way, Arran Road, Ayr Close, Braemar Close, Burns Road, Caithness Road, Caledonian Road, Cromarty Road, Dundee Close, Falkirk Close, Fife Close, Gleneagles Close, Highlands Way, Melrose Close, Montrose Close, Moray Close, Nairn Road, Oban Close, Perth*

Road, *Roxburgh Road*, *Stirling Road*, *Sutherland Way*, Stamford (Lin) (the Scottish names lie on an extensive housing estate by Casterton Road which runs north from the town from *Scotgate* to join the A1 main road to Scotland; the relevance of the latter name to Scotland is in fact disputed, and its real origin may be in *scot*, referring to a local toll or tax).

dukes: *Devon Close*, *Dorset Close*, *Essex Close*, *Kent Close*, *Norfolk Close*, *Rutland Close*, *Westmorland Road*, Redcar (Cle) (compare **counties**, above).

flowers: *Aster Way*, *Azalea Close*, *Begonia Close*, *Dahlia Close*, *Hyacinth Way*, *Iris Close*, *Jasmine Close*, *Lilac Close*, *Lobelia Close*, *Lupin Close*, *Marigold Drive*, *Primrose Drive*, Hinckley (Lei).

folklore: *Allanadale Road*, *Boniface Close*, *Forester Close*, *Friar Tuck Road*, *Greenwood Road*, *Little John Road*, *Longbow Close*, *Maid Marion Road*, *Robin Hood Road*, *Scarlet Road*, *Sherwood Road*, *Tuckswood Lane*, Norwich (a popular choice).

generic identity: *Courtway*, *Crossway*, *Fairway*, *Glebeway*, *Grangeway*, *Greenway*, *Headway*, *Heathway*, *Heronway*, *Kingsway*, *Manorway*, *Parkway*, *Ridgeway*, Woodford Green, northeast London (short cul-de-sacs either side of *Hillside Avenue*; some of these *way* names exist as words in their own right, while others are neologisms devised for naming purposes).

hills: *Cheviot Avenue*, *Chiltern Rise*, *Cleveland Court*, *Malvern Road*, *Mendip Grove*, *Pennine Way*, *The Pentlands*, Peterborough (Cam) (the terrain rises slightly here, northeast of the city centre).

islands: *Arran Close*, *Barra Close*, *Handa Close*, *Lismore Close*, *Manam Close*, *Oronsay Close*, *Rathlin Close*, *Sanday Close*, *Stroma Close*, *Stronsay Close*, *Taransay Close*, *Tiree Close*, *Westray Close*, Hemel Hempstead (Her) (in this case Scottish islands, their names perhaps suggested by *Northend* here).

lakes: *Bassenthwaite Avenue*, *Buttermere Avenue*, *Coniston Avenue*, *Grasmere Avenue*, *Rydal Avenue*, *Thirlmere Avenue*,

Ullswater Avenue, Windermere Road, Birkenhead (Mer) (those of the Lake District, England's best-known group).

military leaders: *Allenby Road, Cavan Crescent, Gort Road, Kitchener Crescent, Marshall Road, Plumer Road, Roberts Road, Wavell Avenue*, Poole (Dor) (in this case all First or Second World War leaders).

musicians: *Arne Close, Beecham Road, Bliss Way, Britten Close, Delius Drive, Elgar Close, Grainger Walk, Handel Walk, Purcell Avenue, Stainer Road, Sullivan Road, Walton Road*, Tonbridge (Ken) (all British or, like Grainger, working in Britain).

ordinal numbers: *First Avenue, Second Avenue, Third Avenue, Fourth Avenue, Fifth Avenue, Sixth Avenue*, Bradford (WYo).

Oxford or Cambridge colleges: see Bootle (Mer) names, above, p. 179.

poets: *Byron Road, Kipling Road, Milton Road, Shakespeare Road, Shelley Road, Spenser Avenue, Tennyson Road, Wordsworth Avenue*, Cheltenham (Glo) (famous for its annual literary festival).

politicians: *Attlee Road, Bevin Road, Keir Hardie Way, Morrison Road, Webbs Road*, Hayes, west London (all Labour leaders in this case; compare **prime ministers**, below).

popular celebrities: *Bannister Close, Bedser Drive, Dimmock Drive, Hapgood Close, Larwood Close, Lilian Board Way, Mary Peters Drive, Matthews Road, Ramsey Close*, Harrow, northwest London (in this instance sportsmen and women: cricketers, footballers and athletes; compare the film actors at Borehamwood, above).

prime ministers: *Beaconsfield Road, Clement Close, Disraeli Close, Gladstone Close, Palmerston Road, Salisbury Street, Stanley Road, Winston Walk*, Ealing, west London (a blend of first names, surnames and titles).

public schools: *Charterhouse Road, Denstone Close, Haileybury Road, Roedean Close, Stonyhurst Road, Wrekin Close*, Liverpool (Mer).

rivers: *Derwent Road, Medway Road, Mersey Road, Severn Road, Trent Road, Wye Road,* Ipswich (Suf) (the housing estate is near the Orwell).

royalty: *Alexandra Road, Andrew Road, Anne Road, Charles Road, Cornwall Road, Edinburgh Road, Edward Road, Elizabeth Road, Essex Road, Gloucester Road, Kings Road, Lancaster Road, Mountbatten Avenue, Norfolk Square, Princes Road, Queen Street, Sandringham Close, Sussex Road, Victoria Road, Windsor Close, Worcester Crescent, York Road,* Stamford (Lin) (a housing estate that has gradually extended; the royal family names are the most recent).

saints: *St Augustine Avenue, St Catherine's Avenue, St Ethelbert Avenue, St Lawrence Avenue, St Margaret's Avenue, St Mildred's Avenue, St Monica's Avenue, St Winifred's Avenue,* Luton (Bed).

Shakespearean characters: *Beatrice Street, Benedict Street, Bianca Street, Celia Street, Falstaff Street, Juliet Street, Macbeth Street, Orlando Street, Othello Street, Portia Street, Romeo Street, Rosalind Street, Viola Street,* Bootle (Mer).

towns: *Camborne Avenue, Dulverton Road, Dunstable Road, Faringdon Avenue, Guildford Road, Kettering Road, Kings Lynn Road, Petersfield Avenue, Tring Gardens* (and many more), Harold Wood, northeast London.

trees: see Staines (Sur) names, above (p. 179).

14. STUDYING STREET NAMES

The reader may well be anxious to make a personal study of local street names, whether on home territory or in a particular town or district.

How should this be done?

The first and obvious requirement is an up-to-date and preferably detailed street plan of the area, and necessarily one with an index. Such street plans are almost always available locally for as little as £1 or even less.

The second requirement, of course, is access to information on 'local history', that is, to publications or documents which, ideally, will record either past forms of a name, where it is an old one, or the circumstances of its bestowal, if more recent. Past forms of a name may not in themselves give the origin of its modern successor, but at least they will indicate how long the name has been in existence and show to what extent it may have been corrupted or altered over the years.

Historic forms of street names, and also of names that no longer exist, may be found in a number of sources, from local rate books to street directories. Kelly's *Directories* are well known examples of the latter, although they date no earlier than 1800. Origins of street names may well be available in a local publication of some kind (see the Bibliography, p. 211, for examples), although not all such publications are comprehensive and some, with the best of intentions, are amateur, and over-reliant on folk etymologies.

For reliable information on the history of street names readers should turn to specialist (but still accessible) works, such as the county volumes published by the English Place-Name Society. The most recent volumes are the most comprehensive, with that for Lincolnshire (Part One, Vol. 58, published in 1985) containing an extensive section on the street names of Lincoln (although only down to 1868) and that for Norfolk (Part One, Vol. 61, published in 1989) doing likewise for Norwich (and including a number of modern names). The *Victoria History of the Counties of England*, in its own many volumes, can equally furnish interesting local information on street names.

Kenneth Cameron, in *English-Place Names* (see Bibliography), devotes a short nine-page chapter to street names which can serve as a useful introduction to the subject. More specifically, it is hardly surprising that the names of London streets should be fairly exhaustively covered, and much of interest will be found in the books on the subject by Gillian Bebbington, Sheila Fairfield and John Field (see Bibliography). Readers curious to discover the now largely lost street names of 17th-century London, together with their contemporary spellings, can do no better than browse through Ogilby and Morgan's *London Survey'd* (see Bibliography), which lists not only streets and lanes but (to quote from its lengthy full title) 'The Courts, Yards, and Alleys, Churches, Halls, and Houses of Note' that were found in them over 300 years ago.

Almost every town in England will have one or more accounts of its history in book form, and such books can also contain information about streets and their names, or at least about local features or prominent personalities who may have given their names to the streets. Some examples of popular local histories of this kind are mentioned in the introduction to the Bibliography, any of whose titles can serve as a first step in the subject.

The general starting point for any serious study of street names, however, should in most cases be the local library, where required maps, plans, lists and background publications will almost certainly be available. Many towns, too, have a local history society of some sort, and the secretary of such a society (whose name can also be found through the library) will doubtless be pleased to help and advise on particular sources. A local museum may similarly hold useful documents recording street names, and the local newspaper office, if not the library, may hold helpful back numbers of a local paper with information about the naming of streets in recent times.

Researchers who desire more detailed or specialist information could well contact their local record or archives office, usually located in a county's administrative centre, such as the Leicestershire Record Office in Leicester, the Lincolnshire Archives Office in Lincoln, or the Hampshire Record Office in Winchester. Such offices hold old town planning records, which in themselves can be a fascinating source of proposed and rejected street names, as well as giving dates of building and names of builders and landowners. Local census returns or registers of electors will provide further information on the inhabitants of individual streets. If the town is an old one, medieval charters for local field names and street names will be found equally valuable. Record and archives offices, and also some of the larger libraries, will also often hold the earliest Ordnance Survey street plans for some of the towns in a county, dating back well over 100 years. The first Ordnance Survey map of London was begun in 1848, on a scale of 5 feet to a mile. Subsequent maps were on different scales, but a further, improved map on the same scale made in 1873 is remarkable for its detail, with names clearly given not only for the smallest of streets and alleys but for individual buildings and even for rooms and offices inside important buildings such as the Bank of England and the Mansion House.

The addresses of county record and archives offices will be found in a source such as Janet Foster and Julia Sheppard's *British Archives* (Macmillan, London, 1982), while useful background information about the exploitation of local archive material is given in F. G. Emmison, *Archives and Local History* (Phillimore, Chichester, 2nd ed., 1978). (Local historians will find the same author's other publications of interest.) For information on local history generally, and explanations of the terminology and types of research involved, the reader is recommended to consult Stephen Friar's *The Batsford Companion to Local History* (see Bibliography).

A little 'field work', or more precisely street work, can also combine business with pleasure, and a walk round the streets of a town or district can be particularly interesting for a study of generic terms. What types of *Terrace* are there, or what appears to constitute a *Close*? Such perambulations can also disclose local street names that are not on the map, either those of streets that have been since renamed (but with the old name still displayed or visible) or those of brand new street names on sites still under construction, for example. In some towns it is possible to find a housing estate where there are names already *in situ* but only rudimentary streets and no actual houses. Chats with elderly local residents may also produce interesting information, as well as colourful or romantic explanations for the names of individual streets. Such 'local lore' should not be disdained, since the names of many streets have evolved today precisely through such folk etymologies.

A field of study in the subject that has as yet been only sporadically pursued is that of street names in fiction. Some authors, such as Dickens, use real street names in their works, although they may take liberties with their actual locations. Others, especially modern popular authors such as Warwick Deeping (see the epigraph at the beginning of this book) and Agatha Christie, devise fictional names for their streets. If a town has a fictional name, its streets are usually fictional as well. A general investigation into the names of streets in fiction could be interesting and rewarding to the literary researcher.

Mention of generic terms is a reminder that the student of street names need not be limited to street name origins. If the medieval language that lies behind several older names appears rather daunting, a study of modern street names, or naming patterns, can be rewarding. Little work, for example, appears to have been done on the relevance of thematic names in a given town. Are they purely arbitrary or locally significant? If the latter, how were they arrived at? A study of local or dialectal generic terms such as *Wynd* and *Twitten* can also bring worthwhile results. Again, the inter-relation between street names and field names, for example, is of interest. Where the historical aspect of street-naming has a greater appeal, a study could be made of 'lost' or changed names. It is even possible to take a particular individual name and establish its frequency and usage in a number of towns. This can be particularly valuable for a name such as *Union Street*, which can have two quite different meanings. Which is the more common?

In short, local or wider research can result in a number of important conclusions about naming processes, and at the same time, given the particular kind of name involved, both contribute to local historical knowledge and make the researcher genuinely 'streetwise'.

APPENDIX

STREET NAME FREQUENCIES

A summary of the current frequency of street names is given below, the aim being to show, in simplified form, which names occur the most often in two different and disparate regions of the country.

Part I contains a count from the 75,000-plus names in the *Complete London Street Atlas*, published in its second edition in 1991 (see Bibliography, p. 215). The range covered by the *Atlas* is wider than its title implies, since its parameters take in not only the whole of Greater London but much of the surrounding area, including such towns as St Albans (Her), Welwyn Garden City (Her) and Harlow (Ess) in the north, Brentwood (Ess) and Tilbury (Ess) in the east, Sevenoaks (Ken), Reigate (Sur) and Guildford (Sur) in the south, and Windsor (Ber) and Slough (Ber) in the west.

Part II is a count from the 30,000-plus names in the *A-Z Manchester Street Atlas*, published in 1991. This includes the whole of the Greater Manchester area, which takes in such towns as Bolton, Bury, Rochdale and Oldham to the north, Stalybridge and Hyde to the east, Stockport, Altrincham and Wilmslow (Che) outside the metropolitan county of Greater Manchester to the south, and Stretford and Irlam to the west.

The count for each name is followed by the generic term (*Road*, *Street*, etc) that occurs most frequently for it. The figure of 323 for *Park* (*Road*) thus means that streets named *Park* were found 323 times, that most of them (actually 85) were *Park Road*, but that there were also many streets named with other generic terms, including *Park Avenue*, *Park Close*, *Park Crescent*, *Park Lane*, *Park Street*, *Park View* and *Park Way*. Where an equal count of generic terms was noted for a name, the one that has the more general application was preferred. This means *Road*, not *Street*, *Street* rather than *Way*, and *Way* rather than *View*. No separate count of generic terms was undertaken.

The frequency count is limited to names with 30 or more occurrences in the Greater London area, and those with 20 or more incidences in Greater Manchester. The count is restricted to those streets (the majority) that have an individual name followed by a generic term, such as *Park Lane*, *Shirley Avenue*, and so on. The count does not include complex names, such as *Grove Park Terrace* or *Broadwater Road North*, since the former type of name is often derivative (*Grove Park* is a street as well as a park) while the latter is repetitive (there is also *Broadwater Road South*, which gives this particular *Broadwater Road* a triple representation). Names consisting of a single word, such as *Greenway*, are taken into account, but names of simplex type preceded by a definite article, such as *The Avenue*, are listed separately (in Part III) as being unique: they mostly consist of a generic term with no defining element.

For the record, it may be noted that *Station Road* is the most common street name in the Greater London area, with 125 occurrences, as against 110 for *Church Road*. It should not be assumed, however, that it is necessarily the most common street name in the country. Its frequency is much lower in Greater Manchester, for instance, and its high rating in the Greater London area results from the fact that the southeast of England has a denser network of railway stations than anywhere else.

The names themselves are a blend of standard words (*Church*, *Park*, *Station* etc) and proper names (*Victoria*, *St John's*, *Tudor*). The latter are mostly royal names and aristocratic titles. High on the list among the latter are the names of Queen *Victoria*, her husband Prince *Albert*, and the popular Princess *Alexandra*, who married the future Edward VII. Titles of dukes and earls, some of them also royal, include *York*, *Cambridge*, *Warwick*, *Gloucester*, *Marlborough*, *Richmond*, *Salisbury*, *Clarence*, *Devonshire*, *Chester*, *Oxford*, *Bedford*, *Albany*, *Clarendon*, *Pembroke*, *Essex*, *Connaught* and *Cumberland*, among others. Most of these derive from place-names (even *Clarence* ultimately comes from *Clare* (Suf)), so in that sense are suitable for adoption as street-names, which are themselves a special sort of place-name in their own right. Royal houses lie behind such names as *Tudor*, *Windsor* and *Lancaster*, while surnames of aristocratic families are represented, among others, by *Grosvenor*, *Stanley*, *Hamilton*, *Russell*, *Spencer*, *Cavendish* and *Howard*. It is worth noting that the royal title of the Prince Regent, commemorated in *Regent Street*, was Prince of Wales, Duke of *Cornwall* (also in the list) and Earl of *Chester*, cited above.

Church dedications are mostly responsible for *St John's*, *St Mary's*, *St George's*, *St Peter's*, *St Andrew's* and similar sanctiferous names. In some instances a popular name may have been directly adopted from a fashionable London street, such as *Grosvenor Square*, *Russell Square* or *Oxford Street*.

In certain cases a particular individual can be discerned (apart from the members of the royal family mentioned). Thus *Gordon* undoubtedly derives from 'Gordon of Khartoum', the Victorian general and administrator, while *Wellington* pays tribute to the victor of Waterloo. *Nightingale* keeps alive the name of the familiar 'Lady of the Lamp', whose first name, *Florence*, was also popular as a street name. *Nelson* preserves the name of the famous admiral, who met his death in 1805 at the Battle of Trafalgar, itself commemorated in *Trafalgar Square*. *Ashley* refers to Anthony Ashley Cooper, Earl of Shaftesbury, who was known as Lord Ashley at the time when *Shaftesbury Avenue* (from his title) was laid out in the mid-19th century. *Stanley*, although an aristocratic surname (of the Earls of Derby), could equally be a memorial to the famous explorer Sir H. M. Stanley, who met up with Livingstone in Africa in 1871. *Carlton*, familiar from *Carlton House Terrace* and the High Tory Carlton Club, derives from Carlton House, built in the early 18th century by Henry Boyle, Lord Carleton (*sic*), who took his title from Carleton (WYo). He was Lord Lieutenant of the West Riding of Yorkshire before he became Chancellor of the Exchequer and Lord Treasurer of Ireland. He died in 1725. *Milton* appears to honour the great 17th-century poet, while

Waverley almost certainly commemorates the novels by Walter Scott, with some streets so named soon after his death in 1832.

A number of proper names are directly from places, not people. *Clifton* refers to the fashionable watering place near (and now part of) Bristol, and *Claremont* derives from Claremont Park, the mansion near Esher (Sur), that was given to Princess Charlotte, the only legitimate child of the Prince Regent, on her all too brief marriage to the future King Leopold I of Belgium in 1816. The house itself took its name from its original owner, Thomas Pelham-Holles, Duke of Newcastle and Marquis of Clare, the latter being the same Suffolk town that gave the title of the Dukes of Clarence. However, *Belmont* does not appear to originate in the place of this name south of Sutton (GtL), although this is itself in Surrey and received its name in the early 18th century at about the same time as *Claremont*, only eight miles away. The literal meaning of both these names, respectively 'bright mount' and 'beautiful mount', may have influenced their popularity as street names. *Beverley*, an attractive name in its own right (hence its adoption as a first name), here probably derives from *Beverley* Brook, a stream that flows through Richmond Park to the Thames.

The names in the first two parts of this survey tell an interesting story, whether mutually compared or contrasted. Manchester, in the northwest, retains in its street names much evidence of its 19th-century reputation as a world centre of manufacture, especially cotton, and of banking. Hence its occupational names, even when derived from surnames (*Walker Street* for fullers of cloth, *Taylor Street* for tailors), and its 29 *Bank* streets. London, in the southeast, has for many years been the focal point of the country's wealth, mainly financial and commercial, but also agricultural. Not for nothing is neighbouring Kent known as the 'Garden of England', and the remarkably high proportion of *Orchard* streets testifies to the fertility of the region.

Despite its status as Britain's capital, Greater London is a region of much wider rural representation than Greater

Manchester, which is still largely urban. Hence London's 'countryside' names. Hence, too, the predominance of *Street* in Manchester as against London's *Road*. It should also be remarked that most of Greater London's *Mill* streets are named for water mills or windmills, mostly grinding corn or flour, whereas in Manchester such streets are usually named for a cotton mill.

Certain local features are reflected in each listing. London predictably has a sizable number (36) of *Thames* streets, while Manchester has no corresponding river name in its top 30. London has 73 *London* streets leading to it, while Manchester has its equivalent in its 30 *Manchester* streets. Manchester has 52 *Moss* streets, relating to the bogs that are found in the region (and indirectly serving as a reminder of Manchester's damp climate). London, by contrast, has no such names, and its *Marsh* streets (the equivalent) are very small in number, and do not even feature in the list. Some of them are almost certainly derived from a personal name, not a topographical feature.

London has (marginally) more *North* streets than South, if only because 'northwards' is the main direction from the southeast. In Manchester, *South* streets, for the same reason, outnumber *North*. London, too, has many more *High* streets than Manchester.

Manchester has half as many *Chapel* streets as *Church*; London only one-sixth the number. Manchester has both *Victoria* and *Albert* among its first half-dozen names; London, although the seat of the monarchy, has neither. It may be assumed that Manchester's 34 *Princess* streets honour Princess Alexandra, mentioned above. Her husband's name may well lie behind the 27 *Edward* streets. By coincidence both regions have the same number (75) of *Stanley* streets, although for Manchester this figure obviously represents a higher proportion. In the latter city the name almost certainly commemorates the Lancashire-born prime minister Lord Stanley (1799-1869). *Derby* street would likewise preserve his title as Earl of Derby. Again, Manchester has more *Oxford* streets than *Cambridge*, unlike

London, where *Cambridge* (82) outranks *Oxford* (53). This is not merely a preference for one university over the other, but presumably indicates that the various Dukes of Cambridge had closer links with London than with Manchester.

Streets named *Mead* are rare in Manchester, perhaps because they smack too much of southern gentility, while London's *Booth* streets can be numbered on the fingers of one hand, unlike Manchester's 24, mostly honouring General Booth, Victorian founder of the Salvation Army. Manchester's *School Lane*, too, comes 20th down the list, as against its 43rd place in Greater London.

Such incidences reflect the differing social backgrounds to the two communities. In names such as *Moss Lane* the linguistic difference is also highlighted.

Both cities, however, have features in common. Both have many *Park* streets, both have more *West* streets than *North*, *South* or *East*, both have their *Ash Grove* and *Prospect Place*, with their distinctive traditional generic terms. Where streets have an identical name and similar frequency in both regions, it may be cautiously assumed that the pattern is likely to be represented nationwide.

APPENDIX

I Greater London

380 Church (Road)
323 Park (Road)
248 Station (Road)
236 High (Street)
202 Manor (Road)
173 Orchard (Road)
150 Green (Lane)
138 Grove (Road)
West (Road)
127 Queens (Road)
126 Elm (Road)
125 North (Road)
124 South (Road)
114 Victoria (Road)
111 Kings (Road)
Meadow (Road)
106 Grange (Road)
100 Priory (Close)
99 Mill (Lane)
97 Beech (Road)
Chestnut (Avenue)
St John's (Road)
95 Willow (Way)
92 York (Road)
90 Cedar (Road)
88 Tudor (Road)
87 Albert (Road)
East (Road)
86 St Mary's (Road)
85 New (Road)
84 Alexandra (Road)
82 Cambridge (Road)
81 Grosvenor (Road)
80 Bridge (Road)
Oak (Road)
Warren (Road)
Windsor (Road)
77 Warwick (Road)
76 Brook (Road)
Princes (Road)
75 Stanley (Road)
Vicarage (Road)
73 London (Road)
72 Farm (Close)
71 Forest (Road)
Gordon (Road)
70 Hamilton (Road)
Hill (Road)
Rectory (Road)
69 Avenue (Road)
68 Gloucester (Road)
Wellington (Road)
67 College (Road)
Milton (Road)
Springfield (Road)
66 Heath (Road)
St George's (Road)
Woodlands (Road)
65 Fairfield (Road)
64 Castle (Road)
Clifton (Road)
Glebe (Road)
Marlborough (Road)
Russell (Road)
63 Chapel (Road)
School (Road)
61 Highfield (Road)
Maple (Road)

Richmond (Road)
Salisbury (Road)
60 Hillside (Road)
Linden (Road)
59 Ash (Grove)
Lancaster (Road)
Mount (Road)
Wood (Lane)
Woodland (Way)
58 Crescent (Road)
Windmill (Road)
57 Carlton (Road)
Clarence (Road)
Spencer (Road)
Woodside (Road)
56 Cavendish (Road)
Lodge (Close)
55 Claremont (Road)
Devonshire (Road)
54 Crown (Road)
Spring (Gardens)
53 Birch (Close)
Chester (Road)
Oxford (Road)
52 Abbey (Road)
Nursery (Road)
51 Bedford (Road)
Beechwood (Avenue)
Beverley (Road)
St Peter's (Road)
50 Howard (Road)
St Andrew's (Road)
49 Albany (Road)
Belmont (Road)
Blenheim (Road)
Nightingale (Road)
48 Ashley (Road)
Clarendon (Road)
Field (Close)
Garden (Close)
Long (Lane)
Sandy (Lane)
47 Cross (Road)
Hazel (Close)
Lincoln (Road)
Pembroke (Road)
Sycamore (Close)
46 Essex (Road)
Hawthorn (Road)
Nelson (Road)
45 Connaught (Road)
Cromwell (Road)
Cumberland (Road)
Holly (Road)
Lansdowne (Road)
Mayfield (Road)
Pine (Close)
Seymour (Road)
Sussex (Road)
44 Bishops (Close)
Buckingham (Road)
Court (Road)
Granville (Road)
Langley (Road)
Mead (Close)
43 Albion (Road)
Bourne (Road)
Laburnum (Grove)
Middle (Road)
42 Bell (Lane)
Broad (Walk)
Lyndhurst (Road)

Market (Place)
Rutland (Road)
Vincent (Road)
Walton (Road)
Western (Road)
41 Acacia (Road)
Dukes (Avenue)
Malvern (Road)
Oakwood (Road)
Waverley (Road)
40 Berkeley (Road)
River (Close)
Rose (Walk)
Trinity (Close)
Valley (Road)
39 Byron (Road)
Cecil (Road)
Central (Avenue)
Hall (Road)
Lawrence (Road)
Osborne (Road)
Riverside (Road)
St Alban's (Road)
St Paul's (Road)
Sutton (Road)
38 Beaconsfield (Road)
Coombe (Road)
Dorset (Road)
Hurst (Road)
Kingswood (Road)
Lime (Grove)
Poplar (Road)
Sutherland (Avenue)
Vale (Road)
37 Belgrave (Road)
Christchurch (Road)
Laurel (Close)
36 Cleveland (Road)
Elizabeth (Avenue)
Kent (Road)
Norfolk (Road)
Temple (Road)
Thames (Street)
Winchester (Road)
Woodfield (Road)
35 Coniston (Road)
Cornwall (Road)
Norman (Road)
Portland (Road)
Prospect (Place)
St James's (Road)
St Margaret's (Road)
Stuart (Road)
Westfield (Road)
Woodstock (Road)
34 Briar (Road)
Brunswick (Road)
Kingsley (Road)
Lavender (Road)
Rowan (Close)
Somerset (Road)
33 Derwent (Road)
Douglas (Road)
Greenwood (Road)
Heathfield (Road)
Kenilworth (Road)
Stanhope (Road)
Tavistock (Road)
Tower (Road)
Vernon (Road)
Westbury (Road)
32 Arundel (Road)

Boundary (Road)
Edward (Road)
Harrow (Road)
Kingston (Road)
Wentworth (Close)

31 Avondale (Road)
Beaumont (Road)
Dale (Road)
Florence (Road)
Jubilee (Road)
St Stephen's (Road)
Shirley (Road)

30 George (Street)
Hanover (Close)
St Michael's (Road)
Sandringham (Road)
Saxon (Road)
Tennyson (Road)
Wilton (Road)

II Greater Manchester

134 Park (Road)
101 Church (Street)
74 Victoria (Street)
60 Beech (Avenue)
62 Green (Lane)
58 Albert (Street)
54 Chapel (Street)
53 Highfield (Road)
52 Moss (Lane)
50 West (Street)
49 Hawthorn (Road)
48 Windsor (Avenue)
47 Ash (Grove)
Stanley (Street)
46 Grosvenor (Street)
45 Wellington (Street)
44 Birch (Avenue)
Manor (Road)
York (Street)
43 Mill (Street)
42 Brook (Street)
Richmond (Road)
Spring (Street)
40 Clifton (Street)
Cross (Street)
Market (Street)
Oak (Street)
South (Street)
Wood (Street)
39 Elm (Grove)
School (Lane)
38 Maple (Avenue)
Meadow (Close)
37 Grange (Road)
36 Hall (Street)
North (Road)
Warwick (Road)
35 Egerton (Street)
Lime (Grove)
Milton (Street)
Pine (Street)
Station (Road)
34 Carlton (Avenue)
Cedar (Road)
High (Street)
Princess (Street)
Stamford (Street)
33 Derby (Street)
George (Street)
John (Street)
Oxford (Street)
32 Woodlands (Avenue)
31 Albion (Street)
Holly (Street)
Peel (Street)
Poplar (Avenue)
Willow (Street)
30 Clarendon (Road)
Dale (Street)
Manchester (Road)
29 Bank (Street)
Derwent (Drive)
Ellesmere (Street)
Nelson (Street)
Rutland (Street)
Vernon (Street)
Wilton (Street)
28 Coniston (Avenue)

Hazel (Avenue)
Laburnum (Avenue)
Lincoln (Avenue)
Queens (Road)
Queen (Street)
27 Chester (Street)
Dean (Street)
Edward (Street)
Howard (Street)
26 Alexandra (Road)
Balmoral (Road)
Cambridge (Street)
Crescent (Road)
Croft (Street)
Garden (Street)
Grove (Street)
Hope (Street)
King (Street)
Rydal (Avenue)
Vale (Road)
25 Grasmere (Road)
Hamilton (Street)
Hill (Street)
Kensington (Avenue)
24 Booth (Street)
Buckley (Street)
Marlborough (Road)
Sycamore (Avenue)
Union (Street)
Vicarage (Road)
Water (Street)
Westminster (Street)
23 Ashfield (Road)
Belmont (Road)
Bridge (Street)
Duke (Street)
East (Road)
Gloucester (Road)
James (Street)
Lancaster (Avenue)
New (Street)
Orchard (Street)
Railway (Street)
Regent (Street)
St Mary's (Road)
22 Barlow (Street)
Belgrave (Street)
Brookfield (Avenue)
Cecil (Street)
Douglas (Street)
Hillside (Avenue)
Mount (Street)
Portland (Street)
Prospect (Place)
Russell (Street)
St John's (Road)
Walker (Street)
21 Buckingham (Road)
Castle (Street)
Cavendish (Road)
Chestnut (Avenue)
Claremont (Road)
Crompton (Street)
Cromwell (Road)
Fir (Street)
Jackson (Street)
Linden (Grove)
Norfolk (Street)
Osborne (Road)
Sandy (Lane)
Taylor (Street)
Thomas (Street)

Turner (Street)
20 Central (Avenue)
Charles (Street)
Chatsworth (Road)
Delamere (Road)
Dorset (Avenue)
Elizabeth (Street)
Gordon (Street)
Hereford (Road)
Kingsley (Avenue)
Langdale (Road)
Lynton (Avenue)
Mayfield (Avenue)
Mellor (Street)
Old (Road)
St George's (Road)
Sherwood (Avenue)
Valley (Road)
Whalley (Road)

III Simplex names

The two lists below represent the most frequently found instances of simplex names, that is, of names formed from a single word (often a common noun) preceded by *The*. It will be seen that the usual generic terms found in street names are in a minority, although *Drive*, *Avenue*, *Crescent*, *Close*, *Rise*, *Street*, *Square*, *Parade*, *Terrace*, *Mews* and *Walk* are present and the first five of these are among the most frequent to occur. The majority of terms are thus adopted from natural features, especially woods and fields (though not *field* itself). Street planners (or namers) no doubt rightly assume that residents will like to live in a street whose name poetically evokes rural beauty on the one hand and a sense of civilization, privacy and security on the other.

The arboreal names undoubtedly predominate, with not only various types of woodland (*Grove*, *Spinney*, *Glade*, *Orchard*, *Paddock*, *Coppice*, *Copse*, and *Woodlands* itself) but different kinds of trees (*Birches*, *Firs*, *Oaks*, *Cedars*, *Pines*, *Poplars*, *Willows*, *Beeches*, *Limes*, *Lindens*, *Rowans*). *Chase* and *Ridings* additionally suggest a tree-scattered or tree-lined terrain where horses are ridden for hunting or exercise. However, in some instances, *Ridings* may well have its original sense, 'land where trees have been cleared', from conjectural Old English *ryding*, to which modern *rid* is related. This is almost certainly the origin of the street of the name in Ealing, west London (see Chapter 4, p. 54). The popular association, however, must be with the *riding* of horses rather than the *ridding* of trees, and *riding* is actually a recognised term for a lane or track through or past a wood where people may ride on horseback. (In *Our Mutual Friend*, Dickens mentions 'a green lane or riding by the river-side'.)

Warren, although suggesting primarily a place where rabbits breed or game is reared, can also imply an area of uncultivated land where trees grow. Verdant landscapes and genteel cultivation are evoked by names such as *Green*, *Croft*, *Gardens*, *Lawns*, *Mead*, *Glebe*, *Park*, *Common*, *Meadows* and

Pastures. The safety and protection afforded by a valley, especially a wooded one, is romantically suggested by *Dell*, *Vale*, *Glen* and *Dene*, while security or desirable solitude is directly alluded to in *Close*, *Retreat* and *Hermitage*. Oddly, the suggestion of exposure to danger is not felt to be present in such names as *Ridgeway*, *Rise*, *Mount* and *Ridge*, although these imply elevated or exposed sites.

Drive, as the most commonly found generic term, suggests almost every desirable feature that a potential resident could wish for: a private, tree-lined road along which one drives, not walks, to a country house or 'stately home' (by inference, the resident's own).

The definite article that precedes the name emphasises the particular street's supposed exclusivity. If you live in *The Close*, then you live in the only one of its kind, a *Close* that has no need of more specific definition. Some of the street names, in fact, resemble those of country houses, so that *The Grange* evokes an opulent farmstead and *The Firs* suggests an elegant mansion surrounded by these lofty and graceful trees.

It will be noticed that five of the top names end in *-way: Ridgeway*, *Broadway*, *Fairway*, *Greenway* and *Causeway*. The first four of these are not properly simplex names, since the *way* is defined by a specific. In *Causeway*, however, the final way has to some extent been influenced by the final syllable of the earlier word *causey*, with the same meaning, so that the term is not thought of as a composite noun. It may have in turn influenced the other terms. *Ridgeway* and *Fairway* are now also unified nouns in their own right, and they are mostly not consciously analysed as 'ridge way' and 'fair way'. While *Broadway* and *Greenway* are still perhaps felt to be composite terms, however, especially when occurring without the definite article, *Broadway* was actually at one time a unified noun (a *broadway*), and has retained something of its original status. Of all the *way* names, too, it is the one that has come to be specifically associated with a street, if only through the famous thoroughfare in New York. *Greenway*, too, although also not a genuine simplex noun, combines two readily associated words, which in combination gave the unified surname *Greenaway*. Hence the inclusion of these five names here as terms of simplex type.

From the lists below it will be seen that, on the whole, Manchester does not readily adopt such names. Where they exist, however, their frequency broadly agrees with those in the Home Counties. The one exception is *Circuit*, as a type of 'ring road' in a residential area. Despite its name, it does not usually represent a circular road or a crescent but is a street that runs along the three straight sides of a square (up, across and down) off another road, which forms the fourth side. It is also significant that *The Crescent* is the most frequent simplex street name in Manchester and district. The planning of residential streets in Victorian times frequently favoured a curving street over a straight one: not only is it aesthetically more pleasing but it often accommodates a greater number of houses.

APPENDIX

A. Occurrences of 5 or more in Greater London area

74 The Drive
70 The Avenue
51 The Green
46 The Crescent
42 The Grove
41 The Close
34 The Chase
27 The Ridgeway
25 The Broadway
20 The Rise
The Spinney
19 The Glade
18 The Fairway
The Mount
17 The Dell
16 The Ridings
15 The Orchard
The Retreat
The Vale
13 The Greenway
The Street
12 The Causeway
The Croft
The Glen
The Square
The Warren
11 The Gardens
The Grange
The Lawns
The Parade
The Terrace
10 The Birches
The Mall
The Mead
The Woodlands
9 The Firs
The Oaks
The Ridge
8 The Glebe
The Mews
The Paddock
The Paddocks
The Park
The Quadrant
7 The Approach
The Common
The Coppice
The Dene
The Meadows
6 The Cedars
The Meads
The Pines
The Poplars
The Willows
5 The Beeches
The Copse
The Hermitage
The Limes
The Lindens
The Link
The Maltings
The Meadway
The Oval
The Pastures
The Rowans
The Walk

B. Occurrences of 5 or more in Greater Manchester area

18	The Crescent
11	The Avenue
10	The Drive
	The Grove
9	The Close
	The Green
5	The Beeches
	The Circuit

SELECT BIBLIOGRAPHY

Many of the titles listed below were actively consulted in the course of work on this book, as were also the 65 volumes of place-name origins arranged by counties that had been published up to 1991 by the English Place-Name Society. There are of course many other publications dealing with the specific origins of street names in England, ranging from books to articles. Details of most of these will be found in Jeffrey Spittal and John Field's comprehensive *A Reader's Guide to the Place-Names of the United Kingdom* (Paul Watkins, Stamford, 1990). Similarly, many books devoted to the history of individual towns will often give the origin of street names. Just four examples of many, and including those consulted for the present book, are John Haddon's *Bath* (Batsford, London, 1973), H. F. Tebbs's *Peterborough* (Oleander Press, Cambridge, 1979), Gerard Young's *A History of Bognor Regis* (Phillimore, Chichester, 1983), and the superb survey by the Royal Commission on Historical Monuments, *The Town of Stamford* (HMSO, London, 1977). Sir Francis Hill's *Medieval Lincoln* (Paul Watkins, Stamford, 1990 [1948]) is an example of a more specialised work which can furnish important information about street names. It, too, was consulted with profit.

Addison, W., *Old Roads of England* (Batsford, London, 1980)

Banks, E., *Streets of Bristol* (Historical Society (Bristol Branch), Bristol, 1981)

Banks, P. R., *The New Penguin Guide to London* (Penguin Books, London, 11th ed., 1990)

Barker, F. and Jackson, P., *The History of London in Maps* (Barrie & Jenkins, London, 1990)

Bebbington, G., *Street Names of London* (Batsford, London, 1972)

Bradshaw, L. D., *Origins of Street Names in the City of Manchester* (Neil Richardson, Manchester, 1985)

Brown, M., *The Street Names of Ilkeston* (Paddock Publications, Castle Donington, 1983)

Cameron, K., *English Place-Names* (Batsford, London, rev. ed., 1988)

Clegg, M., *Streets and Street Names in Ipswich* (Salient Press, Ipswich, 1984)

Copley, G. J., *English Place Names and their Origins* (David & Charles, Newton Abbot, 1968)

Curle, B. R. and Pratt, P. K., *Kensington and Chelsea Street-Names* (Kensington and Chelsea Libraries and Arts Service, London, 1977)

Davenport Adams, W. H., *London Streets* (Henry, London, 1890)

Ekwall, E., *Street-Names of the City of London* (Clarendon Press, Oxford, 1954)

Elliot, C., *Long Distance Trails Almanac* (Lochar Publishing, Moffat, 1992)

Fairfield, S., *The Streets of London* (Macmillan, London, 1983)

Field, J., *English Field Names* (David & Charles, Newton Abbot, 1972)

Field, J., *Place-Names of Great Britain and Ireland* (David & Charles, Newton Abbot, 1980)

Field, J., *Place-Names of Greater London* (Batsford, London, 1980)

Field, J., 'Street-names' *Local Historian* 16 (1984-5), pp. 195-203

Friar, S., *The Batsford Companion to Local History* (Batsford, London, 1991)

Giles, F., 'Why not Charlotte Brontë Circus?' *Sunday Times*, 16 July 1978

Girouard, M., *The English Town* (Yale University Press, New Haven and London, 1990)

Hadfield, J. (ed.), *The New Shell Guide to England* (Rainbird Publishing/Michael Joseph, London, 1981)

Hands, J., *ABC of Street Names in Dacorum* (Dacorum Museum Advisory Service, Hemel Hempstead, 1985)

Harris, C. M., *What's in a Name?* (London Transport/Midas Books, London, 1977)

SELECT BIBLIOGRAPHY

Hibbert, C., (ed.) *The Encyclopædia of Oxford* (Macmillan, London, 1988)

Hindle, B., *Medieval Roads* (Shire Publications, Princes Risborough, 1982)

Hunter, J. *et al.*, *Streets of Windsor and Eton*, rev. by P. Marson (Windsor Local History Publications Group, 1988)

Hutton, K., 'Streets called "gate"' *Local Historian* 8 (1968-9), pp. 288-91

Kendall, H. P., *Streets of Whitby*, rev. by P. Burnett (Whitby Literary and Philosophical Society, Whitby, nd)

Lawrence, B., *Why is it called that? The History and Origins of Maidenhead Street Names* (Royal County of Berkshire Library Service, Maidenhead, 1980)

Leonard, A. G. K., *Stories of Southampton Streets* (Paul Cave Publications, Southampton, 1984)

Leonard, A. G. K., *More Stories of Southampton Streets* (Paul Cave Publications, Southampton, 1989)

Lloyd-Jones, T., *Liverpool Street Names* (privately published, Liverpool, 1981)

London Topographical Society, *The A to Z of Georgian London* (London Topographical Society Publication No 126, Harry Margary, Lympne Castle, 1982)

McGowan, J., *The Origins of the Street Names of Abingdon* (John McGowan, Abingdon, 1988)

McIntyre, A., 'A street by any other name' *Geographical Magazine*, 48 (1975-6), pp. 714, 717

McKenna, J., *Birmingham Street Names* (Birmingham Public Libraries, Birmingham, 1986)

Margary, I. D., *Roman Roads in Britain* (Baker, London, 3rd ed., 1973)

Markham, J., *Streets of Hull* (Highgate Publications, Beverley, 1987)

Markham, J., *Streets of Hedon* (Highgate Publications, Beverley, 1989)

Marriott, P. J., *Oxford Street Names Explained* (privately published, Oxford, 1977)

Martin, J. L. (ed.), *Names of Streets and Places in the Administrative County of London* (London County Council, London, 4th ed., 1955)

Ogilby, J. and Morgan, W., *A Large and Accurate Map of the City of London* (Harry Margary, Lympne Castle, 1976 [1676])

Ogilby, J. and Morgan, W., *London Survey'd: or, An Explanation of the Large Map of London* (Harry Margary, Lympne Castle, 1976 [1677])

Palliser, D. M., 'The medieval street-names of York' *York Historian*, 2 (1978), pp. 2-16

Petersfield Area Historical Society, *Petersfield Place Names* (Petersfield Area Historical Society, Petersfield, 1976)

Rawlings, G. B., *The Streets of London* (Bles, London, 1926)

Rivet, A. L. F. and Smith, C., *The Place-Names of Roman Britain* (Batsford, London, 1979)

Robins, F. W., 'Stories in street names' *Amateur Historian* 1 (1952-4), pp. 278-82

Robshaw, B., 'One for the Road' *The Guardian*, 2 May 1992

Russan, L. and Russan, A., *Historic Streets of London* (Simpkin Marshall, London, 1923)

Smith, A., *City of London Street Names* (David & Charles, Newton Abbot, 1970)

Smith, M., *Stamford Then and Now* (Paul Watkins, Stamford, 1992)

Stokes, H. G., *English Place-Names* (Batsford, London, 1948)

Taylor, C., *Roads and Tracks of Britain* (Dent, London, 1979)

Thorne, J., *Handbook of the Environs of London* (Godfrey Cave Associates, London, 1983 [1876])

Timperley, H. W. and Brill, E., *Ancient Trackways of Wessex* (P. Drinkwater, Shipston-on-Stour, 2nd ed., 1983)

Usherwood, S. and Usherwood, H., *Street Names* (Ginn, London, 1969)

Velvick, S. L. and Willson, E. J., *Street Names of Fulham and Hammersmith* (Fulham and Hammersmith Historical Society, London, 1977)

Wade, C., *Streets of Hampstead* (High Hill Press, London, 1972)

Weinreb, B. and Hibbert, C. (eds), *The London Encyclopædia* (Macmillan, London, 1983)

Wilshere, J., *Loughborough Street-Name Origins* (Leicester Research Department of Chamberlain Music (and Books), Leicester, 1986)

Wittich, J., *Discovering London Street Names* (Shire Publications, Princes Risborough, 2nd ed., 1990)

Zettersten, L., *City Street Names* (Selwyn & Blount, London, 2nd ed., 1925)

A large number of street plans of individual towns and cities were also constantly referred to, the plans themselves dating from the turn of the 20th century to the present day. Additionally, the following four titles were actively at hand for their many street plans:

Automobile Association, *Illustrated Road Book of England and Wales* (Automobile Association, London, 3rd ed., 1963)

Nicholson, *Complete London Street Atlas* (Harper Collins, London, 2nd ed., 1991)

Thompson Local Directories, *Directory of Town Plans in Britain* (Automobile Association, Basingstoke, rev. ed., 1986)

Yellow Pages Guides, *Street Maps of British Towns* (British Telecommunications, London, 1987)

INDEX

This Index is designed to be user-friendly. It is an alphabetical list of all street names in the book about which some kind of information is given, in other words the vast majority. The information may be the origin of the name, its meaning, its history, its early form, its frequency or simply a relevant comment of some kind. A street name may be quoted either for its defining word (usually the first) or for its generic element (usually the second or last). Since information of different kinds can be given about a street name and its component parts, readers should not necessarily expect that the information *they* seek will be found when they turn to the page on which a name appears. But something of significance or interest will in fact be stated about the name on the page given.

The street name is given its full form with the name of the town or (rarely) village where it is located. Towns are given their basic name for brevity, eg Leamington (not Leamington Spa), Stratford (not Stratford-on-Avon). No county name is given except in cases of ambiguity, eg Anson Way, St Ives (Cam), to distinguish this St Ives from the one in Cornwall.

For the sake of simplicity, any street within the area of Greater London is entered as 'London' only. This means that 'London' can locate a street either in the historic City of London, in the wider London as most people understand it, or in the administrative Greater London that includes individual towns such as Kingston upon Thames, Harrow and Croydon. The text will specify the location of the named street more precisely.

Where a street or road is not located within a single town, but runs for several miles across country, it is followed simply by its county name, eg Green Street (Wil). Where the road runs across two or more counties, the county name (usually that of its starting point) is followed by 'etc', eg Icknield Way (Nor, etc). Bridge names in rural districts are also usually identified simply by their county.

INDEX

Street names beginning 'The' are listed under their second word, eg The Avenue appears as Avenue, The. This is for the benefit of readers who, logically, look up a street name under its main word rather than its definite article.

Identically named streets in different towns are given separate entries, eg Westgate, Leeds; Westgate, Mansfield; Westgate, Peterborough. Where the number of towns instanced is greater than five, however, the street name simply appears in capitals, eg FORE STREET. Capitals are also used for a street name considered generally, without reference to a particular town. This mostly applies to names in the Appendix, where frequency figures are given.

Street names in italics are either former names, eg *Blackboy Lane*, Oxford, or historic (early) forms of a name, eg *Flesshewergate*, Nottingham (now Fletcher Gate). A few such early forms are Medieval Latin, eg *nouus vicus*, London. Historic names beginning 'The' are entered like modern names, eg *Markett Place, The*, London. Where such names begin with *Le* or *La*, the Old French equivalent, the order is normal, eg *la Britasche*, Devizes (now The Brittox, indexed as Brittox, The). In a few cases a historic name is quoted in the main text in its modern equivalent as well as in its original form. In such instances both forms are indexed, eg Saturday Market, Nottingham and *le Seterday market*, Nottingham.

Street names in brackets are those that were proposed but were not adopted for one reason or another, eg (Quality Street), Hull.

The alphabetical order takes account of the generic term, so that Gaskell Avenue precedes Gas Street, for example, even though *Gas* in standard dictionaries precedes *Gaskell.*

INDEX

INDEX

INDEX

INDEX

INDEX

INDEX

INDEX

INDEX

INDEX

INDEX

INDEX

INDEX

INDEX

INDEX

INDEX

INDEX